DISCARDED

Development Planning
and
Spatial Structure

Development Planning
and
Spatial Structure

Edited by
ALAN GILBERT

*Lecturer at University College and
the Institute of Latin American Studies,
London*

JOHN WILEY & SONS
London · New York · Sydney · Toronto

Copyright©1976, by John Wiley & Sons, Ltd.
All rights reserved.

Library of Congress Cataloging in Publication Data:
Main entry under title:

Development planning and spatial structure.

    1.   Underdeveloped areas—Economic policy.
2.   Economic development. 3.   Space in economics.
I.   Gilbert, Alan, 1944–
HC59.7.D47   338.91′172′4   75–30804

ISBN 0 471 29904 9

Photosetting by Thomson Press (India) Limited, New Delhi and printed in Great Britain by The Pitman Press Ltd., Bath.

Contributors

Jaya Appalraju, born 1947, Durban, South Africa. Studied at the University of Lund, Fil. Kand. (Geography), 1970; doctoral work, 1970–1973; University College London, Diploma in Development Planning, 1974. Lecturer at University College London, Development Planning Unit since 1974. Consultant to Swedish International Development Association in Ethiopia, 1971. Author of articles on regional development and national urbanization strategies mainly in East Africa.

Piers Blaikie, born 1942, Scotland. Studied at the University of Cambridge, M. A. (Geography), 1965; Ph. D. (Geography), 1970. Lecturer in Geography, University of Reading, 1967–1972; Lecturer, University of East Anglia, School of Development Studies, since 1972; Member of Overseas Development Group, U. E. A., engaged in projects in India since 1966 and Nepal since 1972. Author of *Family planning in India: diffusion and policy*, 1975, and various articles on agricultural organization in India and Morocco, and diffusion and family planning in India.

Don Funnell, born 1945, Exeter. Studied at the University of Cambridge, M. A. (Geography), 1968; Makerere University, Kampala, research student, 1968–1970; University of Sussex, D. Phil. (Geography), 1974. Lecturer in Geography at the University of Sussex, School of African and Asian Studies since 1970; Rockefeller Visiting Fellow to Makerere University, 1971–1972. Author of various papers on small urban centres in the context of a less-developed rural economy.

Alan Gilbert, born 1944, London. Studied at the University of Birmingham, B. Soc. Sc. (Geography), 1965; London School of Economics, Ph. D. (Geography), 1970. Research Fellow, Institute of Latin American Studies, London, 1969–1970; Lecturer at University College and the Institute of Latin American Studies, London, since 1970. Consultant on port expansion in Peru, 1968–1969, the Bogotá Urban Development Study, 1972–1973, and to the Venezuelan

Government, 1975. Author of *Latin American development: a geographical perspective* and various articles on regional planning, industrial location, urban development and health provision in Latin America.

David Goodman, born 1938, Derbyshire. Studied at the London School of Economics, B. Sc. (Economics), 1959; University of California, Berkeley, Ph. D. (Economics), 1967. Research Fellow in Economics, Brookings Institution, 1963–1964; Lecturer in Economics, University of Manchester, 1964–1967; Research Economist, University of California (Berkeley) Brazil Programme, 1967–1969; Project Specialist for the Ford Foundation in Brazil, 1969–1971; Research Fellow at the London School of Economics, 1971–1973; Lecturer at University College and the Institute of Latin American Studies, London, since 1973. Consultant at different times to the World Bank, Organization of American States, and Ministry of Planning, Brazil. Author (with Roberto Cavalcanti) of *Incentivos a industrialização e desenvolvimento do nordeste* (1974) and articles on Northeast Brazil, and employment problems and industrial development in Latin America.

Michael Safier, born 1941, London. Studied at the London School of Economics, B. Sc. (Economics), 1962; doctoral work 1964–1965; University of Chicago, doctoral work, 1965–1966. Research Fellow in Geography, Makerere University, Kampala, 1963–1964; Lecturer at North London Polytechnic, 1967–1968; Research Fellow, Makerere Institute of Social Research, 1968–1970; Lecturer at the Architectural Association, Development Planning Unit, 1970–1971; Lecturer at University College London, Development Planning Unit, 1971–1973 and Research Convenor for the Unit since 1973. Consultant at various times to the United Nations Centre for Housing, Building and Planning, the Government of Uganda, and the World Bank. Author of articles and editor of books on industrial location, urban economics and regional development in East Africa.

Alan Simmons, born 1941, Ontario, Canada. Studied at the University of British Columbia, B. A. (Sociology), 1963; M. A. (Sociology), 1965; Cornell University, Ph. D. (Sociology), 1970. Research Assistant, Dominion Bureau of Statistics, Ottawa; Research Asociate, Cornell University, 1969–1970; Assistant Professor, York University, Department of Sociology and Anthropology, 1970–1974; Visiting Professor, United Nations Demographic Centre for Latin America (CELADE), 1972–1973; Associate Director, Population and Health Sciences, International Research Centre, Ottawa, 1974 to date. Consultant at various dates to the Ford Foundation and the International Development Research Centre, Ottawa. Author of various publications on family planning, migration and social mobility in Latin America.

Anthony Ternent, born 1936, Havana, Cuba. Studied at the University of Miami (Coral Gables), B. A., 1958; the University of Oregon, M. A. (Economics), 1963; Ph. D. (Economics), 1967. Research Fellow and Professor of Economics, University of the Andes, Bogotá, 1963–1965; Visiting Assistant Professor at the University of California (Los Angeles) and the University of Ceará, Brazil, 1966; Professor of Economics at the National University, Bogotá, 1967; Senior Economist at the Pan American Union, 1967; Economist at the World Bank, 1968; Senior Specialist on Urban Development, Organization of American States, 1969–1975; currently consultant on urban economics in Bogotá. Author of various articles on urban economics in Latin America.

Preface

Today, concern about the spatial structure most suited to national development is widespread among governments in the Third World. This is a new phenomenon but one which is growing rapidly in political importance and which everywhere is beginning to affect national economic and social policies. Unfortunately, this feeling is not matched by precise knowledge among planners of what they can and ought to do. This is due to inexperience, to the lack of appropriate advice from planners in developed countries and also to the shortage of concise guides to current planning practice. Consequently, a major aim of this book is to inform planners in developing countries both about the theoretical and academic work which is relevant to their problems and about the experiences of Third World nations which have engaged in some form of spatial planning.

It is also hoped that this book will be useful to academics and will help direct research into 'relevant' areas of this increasingly complex field. Until recently, much research in this area has been motivated more by the wish to confirm theoretical models developed on the basis of First and Second World experiences than by any strong desire to provide guidelines for regional and urban planners in the Third World. The reviews of current research and experience presented in this book, therefore, may help stimulate good research proposals in the future.

What this book certainly does not provide is direct answers to the difficulties of Third World planners. Indeed one of its arguments is that less-developed countries are sufficiently different that what may be appropriate to one must be extensively modified for use in another. Moreover, there are good reasons for suggesting that books which contain instant solutions do a disservice to Third World planning; firstly because good planning demands local answers based on knowledge about local resource availabilities, political aspirations and social conditions, and secondly, because there has been too much blind acceptance of models formulated in the developed countries. The aim of this book is more modest. It presents an informed look at several critical dilemmas facing planners in less-developed countries in the hope of stimulating reasoned planning responses and intelligent research answers to these dilemmas in the future.

In the process of editing this book several people helped me directly and indirectly and I should like to take this opportunity to thank them. My employers gave me sufficient time and facilities to prepare the manuscript and I should like to acknowledge the secretarial help given to me by Annabel Swindells and Rosamund Bushell and the excellent work of Valerie Cawley in preparing the maps. Finally, the contributors are to be congratulated both for their intellectual efforts and for being polite to the editor at all times despite constant provocation.

London.
July 1975

ALAN GILBERT

Contents

Preface	ix
1 Introduction *Alan G. Gilbert*	1
2 Communication, Scale and Change: The Case of India *Piers M. Blaikie*	21
3 Opportunity Space, Migration and Economic Development: A Critical Assessment of Research on Migrant Characteristics and Their Impact on Rural and Urban Communities *Alan B. Simmons*	47
4 The Role of Small Service Centres in Regional and Rural Development: With Special Reference to Eastern Africa *D. C. Funnell*	77
5 Regional Income Disparities and Economic Development: A Critique *Alan G. Gilbert and David E. Goodman*	113
6 Growth-Centre Strategies in Less-Developed Countries *Jaya Appalraju and Michael Safier*	143
7 Urban Concentration and Dispersal: Urban Policies in Latin America *J. Anthony Ternent*	169
Author Index	197
Subject Index	203

Chapter 1

Introduction

Alan G. Gilbert

THE LATE EMERGENCE OF SPATIAL PLANNING IN THE THIRD WORLD

Planning, like foreign aid and the concept of under-development, is primarily a post-war phenomenon. Few developed countries established planning ministries before the middle forties or early fifties and several critical economic and social matters remain outside the sphere of planning even today.[1] Most less-developed countries began to engage in planning still later and are still passing through an experimental stage. Very few countries in the Third World have the long experience of a country like India and in most, national plans only became common during the sixties. As a result, development planning in most parts of the Third World still suffers from growing pains: a lack of skilled practitioners and institutional experience, major deficiencies in data, a lack of political support and often a clear conception of what can or ought to be achieved through planning. In addition, many activities still remain without appropriate authorities to plan them.

Until recently, one such planning lacuna was the field of urban, regional and spatial planning. Before 1960, the number of institutions and agencies concerned with spatial planning in Third World countries could be counted easily. Mexico, Brazil and Colombia had established river-basin commissions along the lines of the Tennessee Valley Authority (TVA), a variety of cities had produced urban structure plans (notably in ex-British colonies where administrators were under the tutellage of British town planners), and certain regions with particularly severe social problems had been provided with development agencies.[2] In general, however, these were *ad hoc* arrangements and were not motivated by any wish to introduce a general programme of regional development and still less by any desire to plan fully the national space economy.

This lack of interest in spatial matters was still more apparent when compared to the situation in most developed countries where even governments (like those of Britain and the United States) which had not responded wholeheartedly to the appeal of 'indicative' national planning had established agencies or introduced policies concerned with urban, local or regional development. In several developed countries such forms of planning had emerged as early as the 1930s. As a response to high levels of regional unemployment highlighted by the world depression, the British Government established the Barlow Commission in 1936 and the United States Government, the TVA in 1941. The main expansion of regional development and dispersal policies, however, followed the Second World War. It was motivated primarily by two related worries. Firstly, there was general concern that certain regions were suffering from slow rates of economic growth and were encountering acute social hardship. Such concern emerged strongly in France, for example, after Gravier's (1947) influential book had highlighted the situation facing France's provincial 'desert'; from the middle fifties the French incorporated a strong regional element into their planning system (Hansen, 1968; Clout, 1972; Bernard, 1970). In Italy, too, the regional element was introduced soon after the end of the war as a realization grew that the country was effectively divided into two disparate halves. In 1950, the famous *Cassa per il Mezzogiorno* was established, which through land reform, industrial investment and the development of infrastructure was to bring the economy of the south into the twentieth century (Lutz, 1962; Holland, 1971; Allen and Maclennan, 1970). In Britain, the Barlow Commission's report (1940) was followed by a series of government initiatives intended to disperse industry, notably through the establishment of industrial estates and tax incentives, and through the discouragement of manufacturing expansion in London (Hall, 1975; Holmans, 1964; Chisholm and Manners, 1971). Even in the United States efforts were made to resolve the problems of high unemployment and low incomes in certain areas; notably through the creation of the Area Redevelopment Administration (ARA) in 1961 (Estall, 1972; Rodwin, 1970).

The second influence encouraging the growth of urban and regional planning in the developed countries was the fear that the further growth of their metropolitan cities would absorb scarce agricultural land and create insoluble environmental, social and economic problems. These concerns were most apparent in Great Britain where strenuous efforts were made to control the expansion of London and, later, the growth of other large conurbations. The designation of 'green belts', the development of 'new towns', and the controls on industrial growth and later office building, reflected these continuing fears. To a lesser extent similar fears motivated planning efforts in France in the sixties.

Both these factors suggest why regional development and urban policies demanded political action in the developed countries. In an environment of growing affluence and increasing belief in future prosperity, lacunae of poverty and underemployment contrasted sharply with labour shortages and congestion in the major cities. In most developed countries, therefore, regional problems were clearly political priorities, which could not be ignored. By contrast, the spatial dilemmas of most Third World countries appeared less clear-cut. It was not that these nations did not contain poor regions, rapidly growing cities, and large regional disparities which demanded political attention. Not only did they face such problems but they suffered from them more severely than most developed countries. But, unlike the situation in the latter, these difficulties swam in a sea of important issues demanding the attention of planners and politicians. Regional unemployment demands resolution in a national environment of full employment, but in the less-developed nations where unemployment and underemployment were widespread, regional problems seemed to be less urgent than the encouragement of rapid industrial and agricultural development. Similarly, the problems of fast growing cities were of a different order and dimension. New York, London and Paris suffered from overcrowding, pollution, poor housing and traffic congestion, and urban planning (with its associated tools, such as physical controls on building, smoke control and traffic planning) was an obvious answer to many of these problems. While several Third World cities suffered from similar difficulties they paled into insignificance in nations where unemployment, poor health and nutrition and wholly inadequate services were so widespread. In any case, before the sixties, the problems of urban size were less critical in the less-developed countries simply because very few cities existed anywhere near the size of those in the developed countries.[3]

Spatial planning was also neglected because of the different ways in which planning was instituted in developed and less-developed countries. In many developed countries, and especially in Britain, local planning emerged before national planning. As a result, there evolved at the local and regional level, professional bodies which demanded solutions and action on spatial issues. In Britain especially, town planners have long constituted an important lobby, reflected in the existence of a national ministry concerned with local government and town and country planning since 1943. By contrast, planning in the less-developed countries normally was introduced from the top down; the establishment of machinery for national planning normally preceded that for regional or urban planning. Even where this was not the case, the national agency invariably was endowed with greater prestige and easier access to key decision makers. This did not guarantee that subnational development and planning would be neglected, but it was inevitable once economists came to dominate national planning agencies. Such supremacy, by a profession which only recently became interested in urban and regional issues and which was frequently hostile to other disciplines' demands for the inclusion of a spatial element, reinforced the political neglect of subnational planning.

Gradually, however, the realization grew that regional development and spatial planning were appropriate concerns for Third World governments. Such a realization grew in part from early efforts at national planning and the discovery that the spatial distribution of investments created as many difficulties and required as much direction as sectoral allocation. This trend was assisted by the work of planners such as Friedmann (1966), Alonso (1969), Rodwin (1970) and Berry (1969) which argued that spatial planning is essential at an early period of economic development rather than being a later and optional planning extra. Without some form of urban and regional policy, opportunities for natural resource development and the integration of marginal social groups might go by default. Unless planners intervened in the distribution of industry, urban centres and transport links at an early stage, they would be unable to prevent the emergence of unwanted 'primate' cities or acute regional income disparities. Friedmann (1966) even went so far as to argue that many less-developed countries required spatial planning more than developed countries. He argued that 'transitional societies are clearly most directly concerned with regional organization, partly because of the spatial shifts involved in moving from an agrarian to an industrial economy, and partly because a large portion of their potential resources are still unutilized' (page 8). By contrast, developed economies should normally be more concerned with overcoming chiefly urban and metropolitan problems.

These intellectual developments were accompanied by other changes in the disciplines of economics, geography and regional science which produced the tools, methods and concepts needed for competent regional and urban planning. The quality of this work had the added virtue that it generated intellectual respect and so eased its adoption by planning authorities and by professionals initially hostile to subnational planning. Certain forms of spatial planning were also encouraged by the active sponsorship of the various United Nations agencies. Having helped in the initial spread of national planning, these agencies began to encourage the wider interpretation of planning and specifically the establishment of spatial and social planning. Advisory missions to Third World planning ministries, research reports by institutions such as UNRISD, seminars for planners and the establishment of centres for regional development all helped the rapid diffusion of this concept.

Finally, the example of certain major 'spatial' projects within less-developed countries encouraged others to introduce regional development programmes. The attempts of the Venezuelan Government to develop the huge mineral and power resources of the Guayana region through the construction of a new urban and industrial complex was widely publicized. The establishment and early 'successes' of Brasília and of SUDENE, the agency created to assist Brazil's major problem area, the Northeast, had a similar effect. Not only were schemes of this kind useful in achieving national, economic and/or political goals but they also promised to assist in the process of regional development (Stöhr, 1975; Gilbert, 1974).

There can be little doubt that this varied activity had a major impact on less-developed countries at least in the institutional sense. Many governments established regional and urban planning divisions within their national planning ministries. Interest in regional development also became manifest in the writing of development plans. The ubiquitous appearance of terms such as 'growth centre', 'homogeneous regions' and 'regional income disparities' testifies to the awareness of the spatial concept. And even if the full regionalization of national development plans and national budgets is still the exception rather than the rule, the trend among national governments is toward rather than away from that development.

Interest in spatial development planning was still more apparent at the regional level. In Latin America the plethora of acronyms associated with regional development and planning agencies is ample evidence of the expansion in offices of this kind. Regional development and planning agencies now exist in most Latin American countries and there are innumerable schemes for encouraging industrial dispersal, reflecting the strong feeling that metropolitan growth should be slowed and poorer regions assisted in their development (Gilbert, 1974). In most parts of Africa and Asia, the vogue for regional and urban development is less fully developed, but is growing in importance. India, of course, is one Asian country with a long tradition of urban planning, and accepted the goal of balanced regional development as early as the Second Development Plan (1956–1960). However, other countries are now incorporating regional and spatial elements into their development plans. In Nigeria a regional planning office was established in the Federal Ministry of Economic Development in 1970, Tanzania has adopted a growth-centre policy and is considering the establishment of a new national capital (see Chapters 4 and 6), and its neighbour Kenya has incorporated a growth-centre strategy into its rural development programme (Taylor, 1974). Similar examples could be cited for many other African and Asian countries.

These practical demonstrations of interest in the spatial concept have been accompanied by a rapid expansion in the literature on urban and regional development in Third World countries. Contributions to this literature have come from various disciplines. Economists such as Alonso (1968, 1969), Lasuén (1969, 1971, 1974), Johnson (1970) and Bergsmann (1970) are extending the work of regional and urban economics into a Third World context. The geographical field has seen the emergence of an extensive literature concerned with the spatial distribution and planning of development. This work includes studies of so-called 'modernization surfaces' (Gould, 1970; Riddell, 1971; Soja, 1968;), systematic studies of selected Third World problems and regions (Odell and Preston, 1973; Gilbert, 1974; Hoyle, 1974; O'Connor, 1971; Dwyer, 1972), studies of regional development experiences in different areas (Stöhr, 1975; Travieso, 1975; Misra, Sundaram and Rao, 1974) and general discussions of the value of existing spatial models for the study of development (Keeble, 1967; Brookfield, 1975; Gilbert, 1970; Connell, 1970). In sociology and anthropology there is a growth of interest in city-rural relationships, in the concept of 'internal colonialism' (González Casanova, 1964–5; Stavenhagen, 1968; Cotler, 1967–8; Cardoso and Faletto, 1969) and in the spatial dimensions of migrant behaviour (Roberts, 1973; Cardona and Simmons, 1975; Balán, Browning and Jelin, 1973). And, finally, in the field of planning, interest in the spatial component has grown dramatically giving rise to texts on regional planning practice (Hilhorst, 1970; Friedmann, 1973) and detailed descriptions and criticisms of regional and urban planning experience in different parts of the world (Miller and Gakenheimer, 1971; Geisse and Hardoy, 1972; Boisier, 1972; Ford Foundation, 1972).

THE DIFFICULTIES FACING SPATIAL PLANNING AND DEVELOPMENT

It is always difficult to generalize about the successes and failures of planning efforts. Planning is not an activity which can be studied scientifically; too many independent variables can affect outcomes, the outcomes themselves may be viewed in vastly different lights according to the value premises adopted, and frequently the diversity of goals and objectives embraced and the failure to express those goals clearly or in terms of specific time periods complicates matters still further. However, this should not deter critics from commenting upon planning efforts either in support of current courses of action or to suggest that all is not well. Both praise and criticism are contained in the following comments. Praise is implicit throughout the discussion because I believe spatial development planning to be a necessary and desirable activity for governments in the Third World. Without it, various opportunities for helping the poorer majority may be missed and the whole development process may be less equitable and humane. But such an outcome is only possible if spatial development planning follows certain directions. And because I believe there is a strong possibility that these directions will not be followed, most of my comments are directed at the less positive aspects of this form of planning. Such an opinion must be highly tentative given the short experience of most less-developed countries in spatial planning. Clearly, too, the same level of criticism cannot apply with equal force to every country, government or agency. Nevertheless, I believe there are several good reasons why current urban, rural and regional development efforts in less-developed countries may prove of little value.

Firstly, urban and regional development policies being a subset of total and national planning activity are bound to suffer from most of the difficulties which confront the latter. These difficulties have been discussed at length in numerous places and there is no need to go into repetitious detail. Suffice to say that criticisms of national planning are diverse. They range across a spectrum from those who argue that effective planning is impossible outside a socialist or communist society (Griffin and Enos, 1970) to those who suggest that national planning is an inappropriate allocation mechanism and that it has 'not served to raise general living standards anywhere' (Bauer, 1971, page 92).

Most critics of planning, however, fall into neither category. They believe that planning is possible in a mixed economy, though often difficult, and that it is the only conceivable way of rationalizing and harmonizing the behaviour of politicians, bureaucrats and the private sector. Even among this majority, however, there is considerable concern about recent tendencies in planning practice. Seers (1972) and Leys and Marris (1971) have criticized the isolation of national planners from grass-roots reality which often makes their plans unworkable. Leys (1972) has argued that planners are insufficiently subtle and politically aware to convince politicians and administrators of the urgency and need for the courses of action they recommend. This inability stems in part from planners' past failures to establish realistic goals. Early efforts at planning were notable for the high degree of optimism with which targets were set. Efforts in India were notorious for this fault (Hanson, 1966; Streeten and Lipton, 1968) but similar difficulties were experienced in the totally different economic environment of Cuba (Seers, 1964; Dumont, 1970). Sometimes this overoptimism led to a lowering of expectations but more often the targets were retained and the means of achieving them radically altered. Thus India's failure to achieve instant 'take-off' through industrialization led to the adoption of new panaceas such as family planning and the 'green revolution' (Lal, 1973), and Cuba's industrialization problems convinced the government to revert to the strategy of exporting as much sugar as possible (Dumont, 1970). Related to this overconfidence and the frequent change in strategies were two further tendencies. Firstly, there was the temptation to 'overplan'; to produce too many plans in too sophisticated a form and without adequate political and public consultation. Secondly, problems were caused through 'pseudoplanning'; where the planner played a 'more or less a ceremonial role in the state machine' (Seers, 1972, page 32) and where the object of the 'game' was more bureaucratic prestige than practical achievement. It may be argued that urban and regional planning is even more subject to these various difficulties if only because it operates at several levels in the government hierarchy. Attempts to regionalize the national budget, for example, may cause political controversy by making explicit previously hidden spatial imbalances in the distribution of resources. The views of national and local planners may differ widely on fundamental issues such as the acceptable level of urban concentration or the introduction and timing of spatial equity policies. Differences of this kind may lead to conflict within the planning bureaucracy and may spill over into political controversy. Even where agreement exists between planners and politicians at the national and the local level other difficulties may arise. A failure by national planners to understand local conditions may make decrees impossible to implement by lower levels of the hierarchy (see Chapter 2). The wide variety of goals adopted by regional agencies such as equalizing regional incomes, developing natural resources, building infrastructure, coordinating local government activities, damping down political opposition, etc. may prove contradictory and will certainly make an evaluation of their efforts highly complex (Gilbert, 1974). The shortage of competent professionals to man urban and regional planning offices may make even the best designed national and regional strategies impossible to organize.

It is not the purpose of this book to explore the detailed problems which may be encountered, although it is an area ripe for future research. Rather we are concerned here with one specific but widespread feature of spatial planning which may lead to unsatisfactory levels of performance in the future.

This is the general intellectual failing to define accurately the problems which planners should resolve and to state clearly the kinds of techniques they require. One major cause of this intellectual failing is the veneration of Third World planners for the planning practices and models of the developed nations. Urban and regional planning, even more than national economic planning, has tended to imitate the efforts of Europe and North America. The reasons for this tendency are not hard to find: the developed nations introduced spatial policies much earlier and therefore have more experience in its application; a great deal of research has been made of these recent planning experiences; and the shortage of professional planners in the less-developed countries, the training of the limited numbers of planners abroad, the absence of ready-made tools for use in less-developed countries all make transfer an appealing alternative.

Unfortunately, it is an alternative which is often unsuccessful. The transfer of methods often occurs with little serious thought and adaptation, even though such methods are being applied to very different social and economic conditions. Methods are often transferred even when they have not been dramatically successful in the country where they were developed. Unsuccessful British experience has often been transferred to India by eager graduates of British planning schools. The Italian experience in the Mezzogiorno was shipped almost wholesale to the Brazilian Northeast (Goodman, 1972). It seems to me that belief in the value of transfer is so widespread that it should be subjected to detailed scrutiny. Several of the following chapters are devoted to themes closely related to this problem, but it would seem useful to consider carefully some of the general difficulties involved in transfer before proceeding with actual examples. Specifically, it seems advisable to examine the two basic assumptions underlying the transfer of techniques, firstly, that the results of spatial planning in developed countries have been successful, and secondly, that successful methods can be applied easily to the conditions in less-developed nations.

HOW EFFECTIVE IS SPATIAL PLANNING IN DEVELOPED COUNTRIES?

Without condemning the valuable contribution which urban and regional planning has made in developed countries, it is not difficult to show that the post-war experience has not been one long continuing success. In the fields of metropolitan planning and employment dispersal especially, Britain, France and the United States have encountered numerous problems. The difficulties facing urban planners in New York and London are legion. Despite sophisticated administrative machines, frequent boundary changes, fiscal adjustments, subsidies, traffic investments, and controls on building it is probable that these cities' problems have never been more severe. Crime rates, traffic congestion, fiscal deficits, and housing conditions are all deteriorating in certain respects even though the *per capita* incomes of the nations in which these cities are located are growing. If Alonso (1971), Richardson (1973) and others are right in suggesting that the problem of these cities is not their size but their organization, then past efforts at urban planning and government must be partially to blame for their current difficulties. Even countries like Japan, which ought to have learnt something from the United States and British experience, do not appear to have gained very much. Planning neglect and the introduction of inappropriate policies seem to be creating major problems in Japanese society. Even major technological success stories such as the 'bullet trains' and the new Tokyo international airport are being attacked on social and environmental grounds.

If urban planning has not been a total success in the large cities of the developed countries, nor have attempts to disperse employment opportunities to regions with high rates of unemployment. Britain by designing special programmes for areas such as South Wales, the northeast and northwest of England, and many parts of Scotland has brought more employment to those regions. But despite this achievement, per capita incomes are still lower and unemployment still higher than the national average and few know at what cost this partial success has been gained. It has not been planned on any clear intellectual grounds, but has been a pragmatic, political response to specific regional problems. It has not even been a consistent response for as Hall (1975) points out it has consisted of a 'kaleidoscopic, often abrupt, series of changes and even reversals in policy' (page 151). Years of government help to Scotland have achieved less than may be accomplished in a few years by the discovery of North Sea oil. The banning of office development in the middle sixties may have helped persuade some companies to move to provincial cities but also led to a major rise in London office rents and the strange phenomenon of 'speculative non-renting'.[4] Similarly in Italy, 'development' of the south achieved some measure of success but was accompanied by a number of less desirable elements. The *Cassa per il Mezzogiorno* established infrastructure in the south, modernized the transport system, compelled state industries to invest sixty percent of their capital budgets in the south, but could not overcome the fundamental 'dualism' of the Italian economy (Holland, 1971; Lutz, 1962). The new industrial growth centres brought the latest technology from the north, but management decisions were still made in Rome, Turin and Milan, the capital intensive plants failed to create jobs in large numbers and these 'cathedrals in the desert' failed to absorb the south's surplus rural labour.

The argument is not that spatial planning and development in the First World has been totally unsuccessful for that is patently untrue; the Swedish and Danish urban planning processes appear to be highly effective, the French are belatedly but successfully attacking their regional problems, and the British 'new town' experience has been worthy of the detailed study it has received. Rather my thesis is that there have been sufficient failures, changes of course, and intellectual inconsistencies to at least pause for thought. The assumption that spatial planning works well in developed countries is only partially true. As a result, the subsequent idea that it can be transferred to other societies in the less-developed world can also only be partially true.

CAN SPATIAL STRATEGIES BE TRANSFERRED EFFECTIVELY?

Even those strategies which have proved effective in the developed nations may be unsuccessful in the Third World. This is not to say that all such methods are inappropriate in less-developed countries, only that the transfer of spatial approaches cannot guarantee good results. This point needs to be underlined since transfer seems to be the principal way in which the 'spatial component' is now being introduced into less-developed nations. Unless national governments, international agencies and planning consultancies realize the difficulties involved, major planning failures are likely to ensue. These difficulties may be summarized under three general headings:

1. Cultural and economic barriers. Without going into detail, it is easy to show that most Third World nations are socially and ethnically more diverse than most developed countries. This diversity tends to place major difficulties in the way of allocation procedures which work effectively in racially homogeneous societies or even in those where one racial group is dominant. What may be a relatively simple operation in many developed countries, such as regionalizing the national budget or reorganizing local government boundaries, may create major political and racial tensions in many less-developed nations. This is especially true in situations where racial conflict or discrimination has been a feature of the recent past. Unfortunately, examples of such conflict are only too easy to cite. They include the Biafran War in Nigeria, the tensions between African and Asian communities in East Africa, between various tribal and religious groups in India, between the old 'regions' of East and West Pakistan, and between Malays and Chinese in Malaysia. Frequently, the artificial national frontiers designed more by the exigencies of war and European greed than by the realities of geography, worsen these situations. Certainly, few countries in the developed world contain so many different ethnic groups with so little empathy or sense of association as do India or Nigeria. By comparison with these countries, the problems of French 'nationalism' in Quebec, the Basques in Spain, the Negro community in the United States and the Welsh, Scots and Irish in Britain pale into insignificance.

Similarly, the economic gulf which divides both regions and groups within the Third World is often of a different dimension to that in Europe or North America. At the height of Britain's regional problems in the thirties, income disparities between the richest and the poorest areas were never as great as those characteristic of less-developed countries today (see Chapter 5). During the twentieth century, the numbers of people near or below subsistence levels in the poor regions of Western Europe were seldom more than 10 percent; in many less-developed countries today 50 percent may fall into this category. Indeed, the very nature of underdevelopment is such that it creates wide disparities between groups and regions.

These differences in income, language, religion, education and history within Third World nations make regional and urban planning a highly complex process. And when additional difficulties such as low levels of national income and the poverty of most national exchequers are recalled, it is clear that planning strategies can only be transferred successfully in favourable circumstances and after very careful thought and modification.

2. Political and administrative barriers. The successful adoption of appropriate imported strategies also may be hindered by political constraints. Throughout the Third World numerous examples exist of good ideas and policies which have been sabotaged by political elites, even though their governments have used spatial rhetoric and have established special regional agencies and programmes. In Venezuela, for example, numerous regional agencies were set up in the wake of the Guayana experiment but were given very little political or financial support (Friedmann, 1966). There are various reasons for such a half-hearted response; for example, the establishment of regional programmes and institutions may be designed principally to appease influential politicians, local public opinion or even foreign economic advisers. Clearly, therefore, the establishment of such a programme does not guarantee its success.

Even when the new machinery is intended to operate effectively, the manner in which politics and administration in some Third World countries operate sometimes distorts its purpose. Good planning requires that politicians lay down objectives, planners produce detailed policies by which those objectives may be implemented, planners and politicians consult the public and affected parties, politicians approve the plan, and then public officials implement it. In the developed countries, even if public consultation has not been developed to a fine art, this is the normal manner in which planning policies are introduced. In most less-developed countries participation is often non-existent and the practice of objective and equitable distribution is often unknown. What in Britain may be allocated in a rational, if not always effective, fashion by public officials, may in certain less-developed countries be allocated through graft and patronage (Achebe, 1966). In certain cases it is even possible for the whole planning process to be altered so that technicians produce the policies, and the politicians only become involved at the important stage of bargaining over the distribution of funds (Daland, 1967). Clearly, the danger with transferring strategies to certain less-developed countries is that politicians may adapt them for their own, and sometimes doubtful, purposes.

Finally, successful transfer may rely upon the implicit acceptance of certain ideological and philosophical goals. Many regional development policies are based on the assumption, for example, that equity is desirable. But, such an assumption may not be held by all governments in less-developed countries; in some places indeed any pretence at such an assumption is refuted by the existence of highly inequitable land-tenure systems. In such circumstances a regional development policy which demands the raising of rural incomes is hardly a viable candidate for transfer. Clearly, therefore, the successful implementation of any transferred strategy is dependent upon political support, and in most cases only those spatial policies which do not threaten the main political elites are candidates for transfer.

3. Modifying imported models and developing indigenous strategies. Unfortunately, there is little sign that these problems are appreciated fully by spatial planners in less-developed countries. One reason for this is that it is often foreign planners and consultants who are engaged in the design and formulation of spatial strategies. Examples of the absurdities which some expatriate 'planners' have perpetrated are too common to need further description (Bauer, 1971; Myrdal, 1970; Illich, 1973; McCallum, 1974), but even where nationals are responsible for the design of spatial programmes the situation may be little improved. One problem is that planners often use inappropriate models and techniques, in the full knowledge that they are unsatisfactory, because nothing else is available. 'Objective' planning requires 'objective' and preferably sophisticated tools for its very existence; where they do not exist they must be invented. The only long-term solution to this difficulty rests in the indigenous development of new techniques and concepts to which I refer below.

However, it is not only the shortage of indigenous methods which is a source of problems. Often it is the blind acceptance of 'foreign' techniques by indigenous planners which is to blame. The international demonstration effect does not only apply to patterns of consumer demand. It is also a feature of the paraphernalia demanded by indigenous governments for the 'development' of their countries. This paraphernalia takes the physical form of sky-scrapers, iron and steel mills, car plants, motorways and nuclear power stations. But it also embraces the desire for intellectual status symbols such as regional planning agencies and development plans, sophisticated techniques such as linear programming and input–output analysis and 'in-phrases' such as growth poles and urban structure plans. The only real solution to this kind of problem is the emergence of independent, socially-responsible and braver planners. How many governments have set up car plants despite their better judgement because planners were not brave enough to argue firmly about the problems the car would bring? How many planners are now wrestling with the problems of traffic congestion and inefficient industrial plants because they were afraid of appearing old fashioned or unpatriotic when the issue of the car plant was discussed? Imported planning strategies, like cigarettes, need will-power to give them up and a great deal of bad planning is due to the lack of such will-power.

A further barrier to the successful transfer of ideas and models from developed countries concerns their testing. Before implementation a model should be fully tested and approved but, by the very nature of spatial planning, this is not always possible. Planners are often called upon to establish a regional planning system even though there is little information on trade flows, rates of migration, government objectives and so on. There is often no time, no data and sometimes no strong political desire to establish whether a new planning system will be effective; it is implemented nonetheless. The possible lesson here is that no planning is usually better than bad planning.

The problem of appropriate transfer is further complicated by the wide adoption of false assumptions about the nature of development and planning. One such assumption is that there exists a universal, unilineal development process along which all countries will one day pass. This assumption, fostered both explicitly and implicitly by several prominent social scientists in the fifties and sixties (Kuznets, 1966; Chenery, 1960; Davis and Hertz, 1954; Lerner, 1958; Clark, 1951), has permitted the transfer of many inappropriate development concepts. Superficial similarities between the situation of less-developed countries today and those of contemporary developed countries in the past are used to justify the application of similar spatial and development strategies. The belief that all less-developed countries will be able to industrialize characterizes much planning practice. The consequent need to concentrate infrastructural investment in large cities, to develop power supplies for industrial expansion, to establish transport systems so as to widen the internal market all depend in part on this assumption. In certain countries such an assumption may well prove correct but is it relevant to all Third World countries? Surely the differences not only in economic potential but also in the aims and aspirations of different governments are wide enough to justify doubts about the validity of this assumption?

Another example of false assumptions affecting the form of development strategy is the common belief that a spatial strategy is identical to a decentralization strategy. Throughout Latin America the first sign of a spatial component in government policy has been the adoption of a scheme to disperse industry or to equalize regional incomes. Such approaches may well be necessary if the poorest regions are to be assisted, but it is only one approach. Many would argue that incomes for the poor may be raised more effectively by encouraging the growth of employment of metropolitan areas (Currie, 1975; Geisse and Coraggio, 1972; Ramírez, 1974). Again the problem has been mechanical imitation rather than sensible transfer. Because Britain, France, the U.S.A. and Italy suffered from regional problems and followed dispersal policies does not mean that Argentina, Chile, Colombia or Mexico should do the same. Even though I believe that a general case can be made against current levels of urban primacy and agglomeration in certain Latin American countries, the assumption that a spatial strategy is identical to a dispersal strategy is untenable.

Questioning the value of transferability is especially vital now that the less-developed countries contain so many planners, politicians and administrators who have been trained abroad. The international diffusion of new planning techniques warrants a study in itself but it would seem a reasonable proposition that the speed with which new planning ideas now diffuse is faster than ever before. The numbers of western trained 'planners' in less-developed countries anxious and able to use sophisticated techniques is growing; the literature conveying the latest methods and ideas is expanding; the number of programmes for training regional and urban planners has increased enormously. In the 1950s a concept like the 'new town' took several years to spread, in the 1960s the concept of the 'growth centre' was absorbed into every national development plan in record time. There are many advantages to be gained from this acceleration in communications. If the new techniques and strategies being diffused are considered carefully by regional planners in each country then no problems will develop. But if the new methods are mechanically absorbed into different social, cultural and economic environments, then there is grave danger for spatial planning in less-developed countries.

The above argument amounts to a simple plea for care and for the development of indigenous approaches. By all means inform the Third World of the techniques, concepts and strategies used in the First and Second Worlds, but let the governments and planners of the less-developed countries apply these innovations only when they have been shown to be appropriate. The mechanical transfer of possibly inappropriate technology is a grave disservice to the poor of the Third World whether implemented by local or foreign planners. So, too, is the current absence of spatial strategies developed by indigenous planners.[5]

This book emerges, then, from a dissatisfaction with our understanding of spatial processes in less-developed countries, with the reluctance of spatial theorists and planners to reject outmoded and inappropriate concepts and techniques, and with recent efforts at regional and urban planning in the Third World. It is not, however, antagonistic either to the generation of theories about spatial process and organization or to the concept of regional and urban planning. Both are clearly required to help overcome the appalling problems which face so many people in Third World countries. Without clear understanding of the spatial processes which govern the movement of people from rural to urban areas, the relations between town and country, and the ways in which certain regions grow and others stagnate, little can be done to help the poor. Spatial planners need to provide appropriate answers to such questions as: whether large metropolitan cities should be permitted to grow, whether new sources of employment should be created in backward regions, and how information on new agricultural techniques can be spread among the rural populace. Answers to questions such as these will never offer a panacea for under-development, but they may highlight and thereby help to ease some of the critical factors impeding more equitable forms of development. At present, however, even this limited goal seems beyond our reach.

The main object of this book is to examine our knowledge of spatial processes and organization in less-developed countries and to suggest issues and approaches which academics and planners might pursue in the future. The book does not try to provide answers to the many spatial dilemmas facing the Third World but does attempt to give a balance sheet of our current assets and liabilities. Each of the following chapters notes the issues on which our knowledge is firmly based, indicates the topics which current work has neglected and examines the approaches which it is hoped will provide interesting research and planning results.

The eight authors, drawn from several social science disciplines to avoid the worst excesses of any single academic approach, have all had experience in the Third World both in research and in planning. Each has tried to make generalizations which may be appropriate to most Third World countries but is aware that what is valid for one less-developed country may be inappropriate elsewhere. For this reason most of the chapters focus on a particular region of the world, sometimes an individual country and sometimes a continent. In this way generalizations are made and explored in the context of a distinct society. Thus many of the problems which Blaikie describes as having affected agricultural-extension and family-planning programmes in India are those which would face similar programmes in other less-developed countries. But his case study also demonstrates many subtle distinctions created by the complexities of Indian society and its government bureaucracy. Hopefully, this counterbalance of generalizations and specific case-studies will underline the point that spatial planners in the Third World must learn from experiences elsewhere but, above all, must remain flexible.

The six chapters are ordered according to the scale of the issues they consider. Starting with those which are concerned primarily with the individual and his relationship with the development process, the chapters move on to the regional scale and the discussion of alternative regional development strategies, and finally to the national scale and the issue of how governments should deal with the growth of giant metropolitan cities.

Blaikie and Simmons both examine processes which directly involve the individual. Blaikie examines the means through which planners have tried to inform individuals about, and persuade them to accept, new styles of life—specifically how to restrict the size of their families and to raise the efficiencies of their farms. He examines the social and administrative barriers which Indian planners have both faced and created, and describes some of the many factors which determine how new ideas are channelled through the complicated social networks of Indian society. Simmons is concerned also with a critical facet of personal life—the decision to migrate. What are the characteristics and motives of migrants, are migrants similar in different Third World countries, and is there any sign that the form of migration changes as economic and social conditions alter through time? He also examines the process of migration from a planning viewpoint and explores the consequences of migration on both the receiving and the departure areas.

Funnell concentrates his attention at the scale of the small town and examines its functions and its potential for development in less-developed countries. If social services are provided what use is made of them by the rural populace? What is the nature of the trading relationship between farmer and small town trader and what conflicts are likely to develop? To what extent can the small town be used to accelerate the pace of economic and social change in rural areas and how specifically do town and country interrelate in East Africa?

Gilbert and Goodman are concerned with planning at the regional scale. They examine the empirical evidence and the assumptions underlying the widely held belief that there is a systematic relationship between levels of per capita income and disparities in regional welfare. They also consider the confusion which exists among regional planners over the desirability of equalizing regional disparities and demonstrate, with examples drawn from Brazil, that such an objective may fail to help and may often worsen, the more critical distribution of personal income.

The regional development theme is also pursued by Appalraju and Safier who examine the application and potential of that ubiquitous technique—the 'growth centre'. To what extent does conceptual imprecision undermine its real value to regional planners and in what different ways have growth-centre policies been implemented in the Third World? Drawing examples from Iran and East Africa they demonstrate how different kinds of growth-centre policy may be appropriate in different social, economic and resource environments.

Finally, Ternent moves to the macro-scale and examines the critical issue of how to treat the giant cities which are emerging in the Third World. At current rates cities like Mexico City and São Paulo could have around 25 million inhabitants by the year 2000. Is this a source of concern and should cities of this size deliberately be discouraged? What evidence is there to support the idea that giant cities are inherently evil or for the counter-argument that they offer the most economic means of achieving rapid rates of economic development? How far do the results of empirical work on this question based on cities in the developed world apply in the conditions of the Third World?

Clearly, these issues do not exhaust the wide range of problems which face spatial planners in the Third World. Nor does this book seek to provide answers to what are extremely complex and intricate issues. What it possibly does offer is a basis for reassessing several critical problems for which we seem to have unsatisfactory solutions. Specifically, it is hoped that it will highlight priorities for research on spatial processes and assist planners in the art of regional development.

NOTES

1. For example the United Kingdom does not employ manpower planning, and prices and incomes are not planned in most West European countries.
2. For example, SUDENE was established in 1959 to deal with the major problem of Brazil's Northeast where constant droughts highlighted the persistent poverty of this region (see Gilbert, 1974)
3. In 1960 no city in the Third World had more than 5 million people and only Calcutta, Bombay, Buenos Aires, Mexico City, Rio de Janeiro and São Paulo more than 3 millions.
4. The dramatic example of the Centre Point office block in Central London and other properties owned by Harry Highams gives support to this view.
5. The need for indigenous approaches should be obvious but its consequence for planning are well illustrated by one isolated concept which has emerged from the Third World—the idea of 'dependency' (Furtado, 1964; Cardoso and Faletto, 1969; Brookfield, 1975). The 'dependency' school interprets underdevelopment not as a persistent and inevitable feature of less-developed socities but as an inheritance of contact with colonial and metropolitan nations. Underdevelopment was caused by the exploitation of the Third World by the metropolitan powers and their agents among the elites of the less-

developed nations. Emerging from this view is a strong distrust of foreign governments and especially their trading policies and investment activities. As a result of this distrust, the 'dependency' school tends to argue strongly for nationalization of foreign enterprises and for the introduction of more equitable development policies. Whether or not readers agree with this interpretation of underdevelopment, this view generates a wholly different respose to development questions. In turn, it is likely to lead to a different form of spatial development policy, viz the experiences and policies of Cuba and Tanzania.

REFERENCES

Achebe, C. (1966) *A man of the people* (London: Heinemann).
Allen, K. and Maclennan, M. C. (1970) *Regional problems and policies in Italy and France* (London: Allen and Unwin).
Alonso, W. (1968) *Industrial location and regional policy in economic development* (Centre for Planning and Development Research, Working Paper No. 74: Berkeley: University of California).
Alonso, W. (1969) 'Urban and regional imbalances in economic development', *Economic Development and Cultural Change*, **17**, 1–14.
Alonso, W. (1971) 'The economics of urban size', *Papers and Proceedings of the Regional Science Association*, **26**, 67–83.
Balán, J., Browning, H. L. and E. Jelín, E. (1973) *Men in a developing society* (Austin: University of Texas Press).
Bauer, P. (1971) *Dissent on development: studies and debates in development economics* (London: Weidenfeld and Nicolson).
Bergsmann, J. (1970) *Brazil: industrialization and trade policies* (London: Oxford U. P.)
Bernard, P. (1970) *Growth poles and growth centres as instruments of regional development and modernization with special reference to Bulgaria and France* (Report No. 70.14 Geneva: UNRISD).
Berry, B. J. L. (1969) 'Relationships between regional economic development and the urban system: the case of Chile', *Tijdschrift voor Economische en Sociale Geografie*, **60**, 283–307.
Boisier, S. (1972) *Polos de desarrollo: hipótesis y políticas—Estudio de Bolivia, Chile by Perú* (Informe No. 72.1 Geneva: UNRISD).
Brookfield, H. (1975) *Interdependent development* (London: Methuen).
Cardona, R. and Simmons, A. (1975) 'Hacía un modelo general de la migración en América Latina', *América Latina: distribución especial de la población*, ed. R. Cardona (Bogotá: CCRP), pp. 3–38.
Cardoso, F. H. and Faletto, E. (1969) *Dependencia y desarrollo en América Latina: ensayo de interpretación sociológica* (Mexico D. F.: Siglo Veintiuno Editores S. A.)
Chenery, H. B. (1960) 'Patterns of industrial growth', *American Economic Review*, **50**, 624–664.
Chisholm, M. and Manners, G. (Eds.) (1971) *Spatial policy problems of the British economy* (Cambridge: Cambridge U. P.).
Clark, C. (1951) *The conditions of economic progress* (London: Macmillan).
Clout, H. D. (1972) *The geography of post-war France: a social and economic approach* (Oxford: Pergamon Press).
Connell, J. (1971) 'The geography of development or the underdevelopment of geography', *Area*, **3**, 259–265.
Cotler, J. (1967–68) 'The mechanics of internal domination and social change in Peru', *Studies in Comparative International Development*, **3**.
Currie, L. (1975) 'The interrelations of urban and national economic planning', *Urban Studies*, **12**, 37–46.

Daland, R. T. (1967) *Brazilian planning and development: politics and administration* (Chapel Hill: University of North Carolina Press).
Davis, K. and Hertz, H. (1954) 'Urbanization and the development of pre-industrial areas', *Economic Development and Cultural Change*, **4**, 6–26.
Dumont, R. (1970) *Cuba: Socialism and development* (New York: The Grove Press).
Dwyer, D. (Ed.) (1972) *The city as a centre of change in Asia*. (Hong Kong: Hong Kong U. P.)
Estall, R. C. (1972) *A modern geography of the United States* (Harmondsworth: Penguin).
Faber, M. and Seers, D. (Eds.) (1972) *The crisis in planning* (Two volumes; London: Chatto and Windus).
Ford Foundation (1972) *Working Papers of the International Urbanization Survey* (19 volumes, New York; The Ford Foundation).
Friedmann, J. P. (1966) *Regional development policy: a case study of Venezuela* (Cambridge, Massachusetts: M. I. T. Press).
Friedmann, J. (1973) *Urbanization, planning and national development* (Beverly Hills: Sage Publications).
Furtado, C. (1964) *Development and underdevelopment* (Berkeley and Los Angeles: University of California Press).
Geisse, G. and Corragio, J. L. (1972) 'Metropolitan areas and national development', *Latin American Urban Research Volume II*, Eds. Geisse, G. and Hardoy, J. E. (Beverly Hills: Sage Publications), pp. 45–59.
Geisse, G. and Hardoy, J. E. (eds.) (1972) *Latin American Urban Research Volume II*, (Beverly Hills: Sage Publications).
Gilbert, A. G. (1971) 'Some thoughts on the "new geography" and the study of "development"', *Area*, **3**, 123–128.
Gilbert, A. G. (1974) *Latin American development: a geographical perspective* (Harmondsworth: Penguin).
González Casanova, P. (1964–65), 'Internal colonialism and national development', *Studies in Comparative International Development*, **1**, 27–37.
Goodman, D. E. (1972) 'Industrial development in the Brazilian Northeast: an interim assessment of the tax credit scheme of article 34/18', *Brazil in the Sixties*, ed. R. J. A. Roett (Nashville: Vanderbilt U. P.), 231–274.
Gould, P. R. (1970) 'Tanzania 1920–63: the spatial impress of the modernization process', *World Politics*, **22**, 149–170.
Gravier, J. F. (1947) *Paris et le desert français* (Paris).
Griffin, K. and Enos, J. L. (1970) *Planning development* (London: Addison-Wesley).
Hall, P. (1975) *Urban and regional planning* (Harmondsworth, Penguin).
Hansen, N. M. (1968) *French regional planning* (Edinburgh: Edinburgh University Press).
Hansen, N. M. (Ed.) (1972) *Growth centres in regional economic development* (The Free Press; London: Macmillan).
Hanson, A. H. (1966) *The process of planning: a study of India's Five-year plans*, 1950–64 (London: Oxford U. P.).
Hilhorst, J. G. M. (1971) *Regional Planning* (Rotterdam: Rotterdam U. P.).
Holland, S. K. (1971) 'Regional under-development in a developed economy: the Italian case', *Regional Studies*, **5**, 71–90.
Holmans, A. E. (1964) 'Industrial Development Certificates and the control of employment in South East England', *Urban Studies*, **1**, 138–152.
Hoyle, B. S. (Ed.) *Spatial aspects of development* (London and New York: John Wiley and Sons).
Illich, I. (1973) *Tools for conviviality* (London: Calder and Boyars Ltd.).
Johnson, E. A. J. (1970) *The organization of space in developing countries* (Cambridge, Massachusetts: Harvard U. P.).
Keeble, D. E. (1967) 'Models of economic development', *Socio-economic models in geography*, eds. R. J. Chorley and P. Haggett (London: Methuen), pp. 243–302.

Kuznets, S. (1966) *Modern economic growth: rate, structure and spread* (New Haven: Yale U. P.).
Lal, D. (1973) *New economic policies for India* (Fabian Research Series 311, London: Fabian Society).
Lasuén, J-R. (1969) 'On growth poles', *Urban Studies*, **6**, 137–161.
Lasuén, J-R. (1971) 'An industrial shift-share analysis 1941–1961', *Regional and Urban Economics*, **1**, 153–220.
Lasuén, J-R. (1974) 'National and urban development', *Symposium on urban development* (Rio de Janeiro: Banco Nacional da Habitaçâo), 89–111.
Lerner, D. (1958) *The passing of traditional society: modernizing the Middle East* (New York: The Free Press).
Leys, C. (1972) 'A new conception of planning?', in Faber and Seers (1968), 56–76.
Leys, C. and Marris, P. (1971) 'Planning and Development', *Development in a divided world*, Eds. D. Seers and L. Joy (Harmondsworth: Penguin), 270–291.
Lutz, V. (1962) *Italy: a study in economic development* (London: Oxford U. P.).
McCallum, J. D. (1974) 'Reflections on foreign planning consultancy', *The Planner*, **60**, 883–888.
Miller, J. P. and Gakenheimer, R. A. (Eds.) (1971) *Latin American Urban Policies and the Social Sciences* (Beverly Hills: Sage Publications).
Misra, R. P., Sundaram, K. V. and Rao, V. S. L. P. (1974) *Regional development planning in India: a new strategy* (Delhi: Vikas Publishing House).
Myrdal, G. (1970) *The challenge of world poverty* (Harmondsworth: Penguin).
O'Connor, A. M. (1971) *The geography of tropical African development* (Oxford: Pergamon Press).
Odell, P. R. and Preston, D. A. (1973) *Economies and societies in Latin America* (London: John Wiley and Sons).
Ramírez, R. (1974) *Planning in a social vacuum: national development and urbanization of Chile* (Royal Town Planning Institute, Overseas Summer School, Exeter, 7–11 September).
Richardson, H. W. (1973) *The economics of urban size* (Westmead, Farnborough and Lexington: Saxon House and Lexington Books).
Riddell, J. B. (1971) *The spatial dynamics of modernization in Sierra Leone: structure, diffusion and response* (Evanston, Illinois: North-Western U. P.).
Roberts, B. R. (1973) *Organizing strangers: poor families in Guatemala City* (Austin: University of Texas Press).
Rodwin, L. (1970) *Nations and cities* (New York: Houghton Mifflin).
Seers, D. (ed.) (1964) *Cuba: the economic and social revolution* (Chapel Hill: University of North Carolina Press).
Seers, D. (1972) 'The prevalence of pseudo-planning' in Faber and Seers (1968), 19–34.
Soja, E. W. (1968) *The geography of modernization in Kenya: a spatial analysis of social, economic and political change* (Syracuse: Syracuse U. P.).
Stavenhagen, R. (1968) 'Seven fallacies about Latin America', *Latin America: reform or revolution?* Eds. J. Petras and M. Zeitlin (Greenwich, Connecticut: Fawcett), 13–31.
Stöhr, W. (1975) *Regional development experiences and prospects in Latin America* (Paris and The Hague: Mouton).
Streeten, P. and Lipton, M. (Eds.) (1968) *The crisis in Indian planning* (London: Oxford U. P.).
Taylor, D. R. F. (1974) 'Spatial aspects of Kenya's rural development strategy', *Spatial aspects of development*, Ed. B. S. Hoyle (London: John Wiley and Sons), pp. 167–188.
Travieso, F. (1975) *Ciudad, región y subdesarrollo* (Caracas: Fondo Editorial Común).

Chapter 2

Communication, Scale, and Change: The Case of India

Piers M. Blaikie

OVERVIEW AND APPROACH

Current geographical literature has distinguished two distinct (although often simultaneous) types of diffusion: that which spreads from neighbour to neighbour in a spatial sense, and that which occurs from one level of a hierarchical organization to other units at that level and proceeds from thence to and among other units at a lower level in the same organization. We shall call the first the neighbourhood effect and the second the hierarchy effect. The former implies local contagion leading to space-filling by the phenomenon being diffused, and the latter to a transmission of the phenomenon across larger distances, leap-frogging from one unit at a particular level in a hierarchy (which in many geographical studies would be towns of a certain size) to another. While this distinction has yielded quite valuable insights into the spatial structure of diffusion it has contributed less to the study of process. In particular it has not yet been used to show that the two forms of diffusion are articulated by very different and sometimes opposing forces of social change.

It is the purpose of this chapter to explore in the Indian context the effect of the social structure upon the ways in which change is channelled: its spatial form and the feedback effect in terms of the new forms which change takes, its extent, and its effect upon the social structure at a later point. The particular issue which this analysis emphasizes is the effect of selective diffusion of new practices upon interfamily and interregional welfare and incomes. Firstly, the process of hierarchical diffusion is discussed in general political terms, to show specifically how government directives percolate down the executive hierarchy from New Delhi to the most remote villages. The filters which operate within and between levels of the hierarchy are as much a part of the political process as the formulation at a higher level of the directives themselves. To claim that the way in which directives are carried out constitutes a diffusion process of the hierarchical type is perhaps to stretch the term too far, although to make a rigorous distinction between the diffusion which is subject to stochastic processes but which tends to be transmitted along existing social networks on the one hand, and a diffusion of directives down an administrative hierarchy on the other, is difficult. While the informal social network has distinct spatial biases (whether it is interurban, and therefore perhaps hierarchical, or local and between neighbours) the administrative structure is usually designed to overcome these biases—although resulting diffusion outcomes may not. Except for certain planned variations in administrative 'density' over space (the Intensive Agricultural Area Programme or the Family Planning Intensive District Programme are two Indian examples), it is the generally implied objective to locate administration at all levels of its hierarchies by some rule-of-thumb algorithm related to numbers of various target populations served per administrative unit. Hence the administrative hierarchy is designed to bring change programmes to the people by overcoming spatial friction. However certain political and organizational influences affect the flow of directives down the hierarchy and finally their execution in the field, such that many of the attributes of what can legitimately be described as a diffusion process can be found to operate within the administration. Next, hierarchical diffusion within the private sector is briefly surveyed. Business networks and central places are discussed, with special mention of how existing rural-urban networks and the nature of transactions flowing along them strongly affect the way new economic opportunities and other innovations are taken up. The latter in turn affects the distribution of income in a manner which ensures the exclusion of larger sectors of the population from the benefits of further technological change.

Secondly, the neighbourhood effect is taken for discussion. The conditions in which this type of diffusion operates in India are quite different from elsewhere. Spatial immobility and caste segregation tend to make diffusion slow and sticky, although the relationship between the characteristic of the item being diffused and the particular nature of the articulation of caste practices has to be carefully studied in each case. The spatial leap-frogging of the diffused item common in Europe and North America is quite absent here. Although rural and poor urban social relationships may be predominantly vertical and involve patronage and dependency, it will be claimed that innovation diffusion is predominantly lateral and that the 'trickle-down' effect in the Indian case is largely a figment of the rural sociologists' imagination and a case of political wishful thinking. The neighbourhood effect can be seen to be highly selective—for no more complicated a reason than neighbours themselves are selective. In detail, and at the disaggregated level of study, the neighbourhood effect is seen to occur only when diffusion is predominantly lateral and not vertical (through 'gate-keepers' or from rich to poor, literate to illiterate, the well-travelled to the immobile peasant). Implications for extension policy and income distribution are discussed. Throughout the discussion, a locational focus upon one of the poorest tracts of one of the poorest states will be taken—

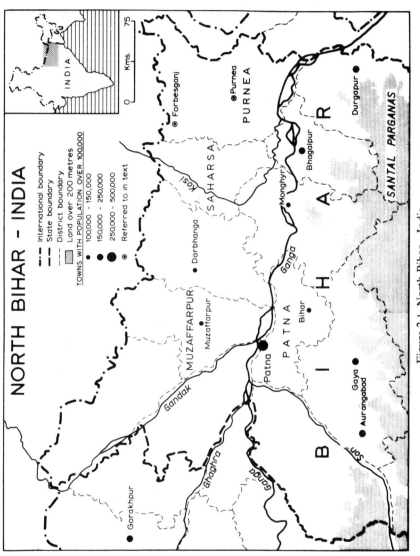

Figure 2.1 North Bihar—India

Purnea and Muzaffarpur Districts of Bihar state, although at higher levels of generality, all Bihar and India will be used for supporting evidence for the argument (Figure 2.1).[1] Within these two districts, the Family Planning Programme and the Small Farmers Development Agency will be singled out for more detailed treatment.

HIERARCHICAL DIFFUSION AND GOVERNMENT PROGRAMMES

In this section, the basic theme is to examine the relationship between nationally stated objectives and the actual work carried out by government servants to achieve them. More than a few nationally acclaimed movements in India have had a scarcely perceptible effect in the field and have become acknowledged failures. The Community Development Programme and Land Reform Programmes are two examples and in recent years the two programmes reviewed in this chapter, the Family Planning Programme and the Small Farmers Development Agency (limited to an area in north Bihar State) are in danger of turning out the same way. The outline design of these programmes arises from many different motives, some popularly political and some (particularly the Family Planning Programme, which was and still is unpopular with all the main political parties) from a sober and rational evaluation of economic and social choices. Land reform, for example, was carried through Legislative Assemblies in grand rhetorical manner but largely failed to fulfil its purpose in the field. The major reasons for this failure are suggested in the following quotations:-

'To democratize the villages without altering property relationships is simply absurd' (Barrington Moore, quoted by Wood, 1975, page 394).

'An observer in Purnea is not surprised by owners who view reform measures as if they were not meant to be enforced and utilize every loophole to their advantage' (Ladejinsky, 1969, page 1).

'... to sum up, only a small percentage (about 5 percent) of the sharecroppers have adopted the new technology to a limited extent. The major handicaps against their adoption of the new technology are (i) the security of their terms (exacerbated by landlords wishing to dispossess tenants of their rights so that they (the landlords) can modernise their land by availing themselves of the new technology—author's insert), (ii) Lack of credit facilities and non-availability of inputs, and (iii) social backwardness' (Prasad, 1967, Chapter 4).

The implications of the failure of much of the Land Reform legislation greatly affects the diffusion of innovations, as will be discussed later.

The Family Planning Programme had no such socialist rhetoric—it was unpopular in many government quarters from the start. Since the 1961 census showed an alarming decennial population growth, the 'clinical' approach was abandoned for an 'extension' approach with a vastly increased network of outlets for the supply of information and supplies. Once the Family Planning Programme had taken firm root as national policy in New Delhi, a whole series of directives had to be issued from the Ministry of Health to state governments. It is the content of these directives and their execution in the field that is of special interest here. Seventy-five directives from the Bihar State Government to the District Family Planning Bureau of the two Districts of Purnea and Muzaffarpur were studied. The collection of all the major statistics for each block in these districts with extensive field interviews of personnel and the public enabled a realistic evaluation of these directives to be made. Basically, there were found to be three types of directive, each with a different diffusion pattern.

1. Objectives and related directives that are almost impossible to carry out, whether physically or administratively, and must be accorded the status of political or administrative rhetoric. One such example is illustrative 'The reduction in the annual (population) growth rate to 1.5 percent per year by 1978–1979 is sought to be reached by extending family planning education to motivate the people in favour of the small family norm and providing services and supplies through mobile and static units as near the homes of the people as possible so that motivated people can readily avail of them' (Letter from Additional Director of Health Services (FP) to District Family Planning Officers, Intensive Districts No. 558 (FP), 11.3.69). In effect, such statements are little more than wishful thinking, but are used, together with prominent reporting in the press, in family planning circulars, in journals of population seminars and of international bodies such as WHO and ILO and in quotations about the dangers of projected current population growth from well-known persons, to give the impression that the programme is in good heart. Although there is much published which is objective and statistically sound, there is much published information (often in the form of directives following ambitious objectives) which amount to no more than public relations.

2. The second type of non-implementable directive is that which is issued at a lower level, either from the state to the district, or the district to the block. For example, the detailed locational policy mentioned above involves the drawing of maps of the area to be served by the programme, registering eligible couples (the target population which consists of all married couples from 15–44 years), canvassing village leaders, marking out areas of responsibility for each programme worker, etc. Of course this sort of directive has to be issued in order to get the programme working at all, but its measure of success depends upon the existing infrastructure in the area. The experience in North Bihar is undoubtedly that such detailed preparation (quite apart from execution) is quite out of the question with existing staff shortages. A plea for the regional planning of the family planning programme, matching objectives realistically with resources (either existing or those which can be planned for with some prospect of fulfilment) has already been made (Blaikie, 1972 and 1975). The basic problem is that there is a lack of information feedback about conditions

at the district level and below, with the result that planners are unable or unwilling to vary resources between regions. In poor, backward areas of India, because of a whole complex of co-linear variables on both the demand and supply side of the programme, many directives do not get diffused nor carried out at all. A number of guidelines laid down by the Ministry of Health in New Delhi, for example, seem to be considerably distorted in the directives from the Secretariat at Patna to Districts in Bihar. The unpalatable probability is that the Departments of Family Planning at the state capital of Bihar suffers from the malady of all government in Bihar—astonishing impermanence, widespread and acknowledged corruption and a lack of purpose and continuity, and in particular from the low status accorded to it by administrators and politicians. This results in its being led by elderly doctors, who are often just about to retire from government service anyway, and its being understaffed in most of its offices. In this case it can be ventured that distortions of directives occur at the state level simply because of administrative inadequacy. At a lower level, the transmission of directives from district to block and below is often faulty because of the sheer impossibility of some of them, and because of a lack of understanding and sympathy by main officers with the objectives of the programme. This lack of understanding can certainly not be the fault of the Ministry which issues a multiplicity of directives, exhortations, guidelines, booklets, statistical tables, etc. but of the system of training of the doctors working on the programme. The syllabus of most medical colleges is based on the British style of forty years ago. The doctor is trained as a member of a healing élite, with strong emphasis on curative medicine (King, 1972). As such he is in no way interested in family planning, which requires a strong extension element to the population he treats and considerable administrative duties. Hence many doctors at the local level (the urban hospitals, subdivisional hospitals, and Primary Health Centres at block level) simply do not pass on directives at all to other programme workers in their charge. Fifteen mandatory monthly coordination meetings at subdivisional level in Purnea and Muzaffarpur Districts at which the author was present were attended by an average of only 63 percent of Block Medical Officers and therefore many *ad hoc* orders were not passed on to the block level. Although these doctors often perform a few vasectomy operations to keep out of trouble, they often disapprove of the whole concept on moral grounds, or simply do not consider it to be their duty to extend services and information. Thus, at all levels of the hierarchy, information is filtered, distorted, or is not transmitted in any form because some branches of the hierarchy at each level happen to be ill-suited to administer the programme and see that directives are carried out. Some of these 'branches' (state, division, district, block) suffer from such a serious combination of adverse conditions, that directives are hardly carried out at all over large areas knowledge of family planning is practically nil. Casual inspection of maps of family planning adoption shows frequently a step-like gradient from one administrative area (controlled by one branch of the hierarchy) to another.

3. Perhaps the greatest barrier to neighbourhood diffusion is the way in

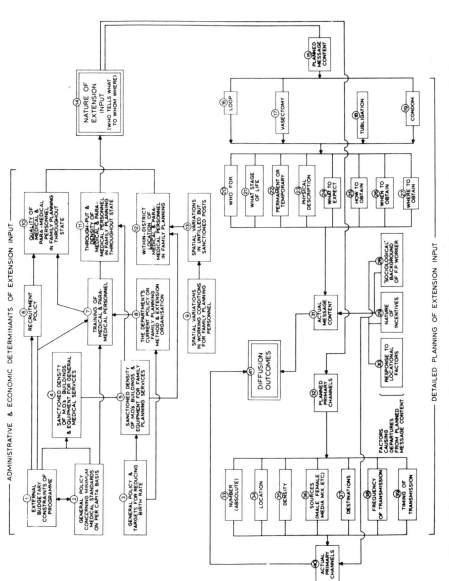

Figure 2.2 General planning outline for extension input of the family planning programme of India

which directives to 'grass-roots' programme workers are distorted. This type of directive is defined not by its content (as 1 and 2), but by its typical level in the hierarchy. Figure 2.2 shows the general way in which the programme is administered in the field, and in particular the extension part of the programme. Directives are received by district level officers which specify the channels and content of extension (subject to the losses of information and addition of 'noise' mentioned in 1 and 2 above).

Some of the ways in which both planned channels and planned information content are miscarried are summarized here (Blaikie, 1974). The planned message content (disaggregated into items of information, boxes 20–27) is often seriously distorted and abbreviated because of the ways in which programme workers have to operate. Each of them is target-chasing (numbers of vasectomies, loop insertions, tubeligations, or condoms distributed per month or year). Since it is easier to measure the number of patients who are 'motivated' and who agree to a form of family planning, then to evaluate more general extension work, each level of the hierarchy is carefully watching the levels below which it controls, anxious that the aggregate target appropriate to each level will not fall so spectacularly behind that blame might fall upon the individuals at that level and below. Hence the emphasis is upon 'capturing' patients, and in order to do this, the incentive money for the operation or loop insertion is emphasized (18 rupees for tubeligation, 15 rupees for vasectomy and 6 rupees for loop insertion in Bihar,[2] but generally more in other states) and the minimum information concerning where and when the service can be obtained. As the 'canvasser' also gets a smaller sum for 'bringing in' a would-be family planner, it is easy to see how the government is in fact asking programme workers to 'sell' information without any means of quality control. Since non-government people can also canvass, there has arisen the phenomenon of the 'professional' canvasser. In Bihar these are often, slightly disreputable people who haunt railway stations and markets, wherever a crowd of itinerant labourers or others short of ready cash congregate. Frequently, the messages they impart are grossly distorted, particularly with regard to the physical description of the family planning method (23), and what to expect (24). This is to be expected when the government contracts out 'extension work' to any private individual for personal gain.

Planned Primary Channels (32) are not carried into effect for similar reasons. Many programme workers are urban-orientated people who have only taken the job because they could not find a job in the town, and consequently they intensely dislike staying in villages. Although officially, extension educators are supposed to make twenty 'night-halts' in villages per month (ensuring a seasonal spatial penetration of family planning messages), a strong distance-decay effect characterizes visits from block headquarters. Furthermore the extension educator often feels exasperated by the poor's response to family planning (usually solid objection or distrust for very sound functional reasons, discussed below) and therefore prefers to talk to the educated in the villages, thus biasing the planned distribution of destinations (37). Because of a lack of supervision of work at all levels, the actual number (33), location (34) and density (35) of channels in the field frequently reflect a minimum commitment sufficient to keep out of trouble. (It must be emphasized that elsewhere in India, these depressing remarks are not relevant, as in a number of states the programme is very creditably managed, notably in Maharashtra, Haryana, Punjab and Mysore.)

The Small Farmers Development Agency offers a useful contrast in the type of network through which directives move, in the service it offers, and in its target population. The programme covers that part of Purnea District and its neighbourhood Saharsa District which is under command from the eastern part of the Kosi irrigation system (and very small areas outside it), and so is only a 'local' programme, as distinct from the family planning programme which is national in every respect. It had been recognized for some time that most of the essential inputs for an improved agriculture were being purchased by large farmers. In the case of tubewells, fertilizers, pesticides, high-yielding varieties of seeds, and credit, small farmers had neither the capital, the land to mortgage to raise it, nor in some cases plots large enough to make a tubewell of government specified dimensions worthwhile (Clay, 1974). 'It is not, however, the new technology which is the primary cause of the accentuated imbalances in the countryside. It is not the fault of the new technology that the credit service does not serve those for whom it was originally intended; that the extension services are not living up to expectations, that the *panchayats* (elected village government bodies, with considerable powers of distribution of credit and inputs) are political rather than development bodies; that security of tenure is the luxury of the few; that rents are exorbitant; that ceilings on agricultural land are notional; that for the greater part tenurial legislation is deliberately miscarried; or that wage scales are hardly sufficient to keep body and soul together. These are man-made institutional inequalities' (Ladejinsky, 1969, page 15). The Planning Commission, New Delhi, recognizing these difficulties encountered by small farmers, commissioned a study in 1969 entitled *The Problems of Small Farmers in Kosi Area*. On the basis of this document which attempted to identify the numbers of 'small' farmers (2.5–5.0 acres) in the area, their present levels of usage of new inputs, their present and future requirements, and some of the major problems which they faced, the Small Farmers Development Agency (SFDA) was set up in 1970. The agency attempted to coordinate the provision of facilities for the small farmers (primarily credit) which was (and is) provided from a multiplicity of normal block and additional Intensive Agricultural Area Programme (IAAP) outlets. It was instructed to concentrate on farmers with holdings between 2.5 and 5.0 acres (these were assumed to be the two limits which bounded 'viability' at the lower level and a 'small farmer' at the upper). The SFDA is essentially a co-ordinating agency which attempts to identify small farmers, make lists of them available to other agencies and to extend through existing agencies short-term credit for the purchase of fertilizers and high-yielding seed, medium-term credit for the purchase of bullocks or small-scale capital equipment and for land improvement, and long-term credit for the purchase of large tubewells, electric pumpsets, tractors and for major land improvements.

The SFDA suffers from similar problems to those of the family planning programme. First of all, the registering of small farmers had to be carried out by Village Level Workers (VLW) who should have been in the best position to know the whereabouts of small farmers in the villages of their area (as they are local persons living in or near the villages they serve). However, many small farmers were omitted from the registers, mostly through a lack of methodical search by VLWs but also through wilful exclusion. The numbers excluded are difficult to assess, but in the blocks surveyed the Project Executive Officers and Block Development Officers estimated that between twenty-five and thirty percent of small farmers were excluded. Furthermore those excluded found it

```
                        Short-term credit
                          1.00 (63)
   never heard of facility                 heard of facility
   C. 14 (9)                                0.86(54)

                    mukhiyaa   VLW      PEO      Coop.    Other
                                                 Sec.     farmers
                    0.10       0.43     0.24     0.04     0.04
                    (6)        (27)     (15)     (3)      (3)

   never heard of procedure              heard of procedure
   0.03 (3)                               0.81(51)
                    mukhiyaa   VLW      PEO      Coop.    Other
                                                 Sec.     farmers
                    0.08       0.41     0.22     0.04     0.04
                    (5)        (26)     (14)     (3)      (3)

   did not apply for                    applied for STC
   STC 0.43 (27)                         0.38(24)
no    other     too         not
water 0.03(2)   complicated  timely
0.16            0.16(10)     0.08(5)
                                        (first-time
                                         successful delivery
                                         0.14 (9))
                             unsuccessful
                             0.24(15)
   no follow-up   follow-up
   0.03(2)        0.13(13)
                             delivery
                             0.29(18)
   still unsuccessful   bribe
   –no loan 0.03(2)     0.10(6)
                             preliminary technical
                             advice 0.16(10)
   no preliminary advice
   0.13(2)                   follow-up advice 0.02(1)
   no follow-up advice
   0.14(9)
```

Figure 2.3 Adoption outcomes for short-term credit

Medium-term credit
1.00 (63)

	mukhiyaa	VLW	PEO	Coop. Sec.	Other farmers
never heard of facility 0.49(31)			heard of facility 0.51 (32)		
	0.11(7)	0.29(18)	0.05(3)	0.02(1)	0.05(3)
never heard of procedure 0.25 (16)			heard or procedure 0.25(16)		
	mukhiyaa	VLW	PEO	Coop. Sec.	Other farmers
	0.05(3)	0.16(10)	0.02(1)	0	0.03(2)

did not apply for STC 0.03(2)
applied for MTC 0.22(14)

too costly 0.03(2)	unsuccessful 0.06(5)			delivery 0.18(9)	
no follow-up 0.02(1)	follow-up 0.06(4)				
still unsuccessful no loan 0.05(3)	bribe 0.02(1)				

Figure 2.4 Adoption outcomes for medium-term credit

very difficult to get credit, and usually (for the exceptionally tenacious small farmer) it took more than a year and possibly two to have himself entered into the register.

Figures 2.3, 2.4, and 2.5 show the proportion and actual numbers of small farmers in a sample taken in Purnea District from four neighbouring blocks. In this section, the governmental role and its shortcomings in providing ex-

Long-term credit
1.00(63)

	mukhiyaa	VLW	PEO	Coop. Sec.	Other farmers
never heard of facility 0.81 (51)			heard of facility 0.19(12)		
	0.21(1)	0.11(7)		0	0.64(4)
never heard of procedure 0.03(2)			heard of procedure 0.16(10)		
	mukhiyaa	VLW	PEO	Coop. Sec.	Other farmers
	0	0.11(7)	0	0	0.04(3)

did not apply for LTC 0.11(7)
applied for LTC 0.04(3)

too expensive and/or fragmented land 0.11(7)	unsuccessful 0.03(2)				
	follow-up 0.03(2)				
still unsuccessful 0.03(2)	delivery 0.02(1)				

Figure 2.5 Adoption outcomes for long-term credit

tension loans and inputs is described while the diffusion of the government programme as it is carried out in the field amongst its target population is discussed in section three. Taking Figure 2.3 first, showing the outcome of the short-term credit programme, it is clear that most of the farmers had heard that the facility existed, with the VLW and PEO (the Project Executive Officer who tours villages in his block and holds meetings in each *panchayat* of which these are about one hundred or so in each block) dominant as sources of information. The *mukhiyaa* or elected leader of the *gaon panchayat* (village council of five members) is unimportant as an extension agent here. Most of those who had heard of the facility had also heard of the procedure required to avail themselves of it, but problems begin to arise once the farmer has applied for credit. Rumours about costly follow-up visits, long waits outside offices, and a few bribes had in fact already found their way back into the information pool such that fifteen small farmers did not think it worth bothering about in the first place. Medium-term credit has a poorer showing with a serious extension shortfall (even more serious than the proportions indicate when *all* farmers, including those not registered, are included). The procedure for medium-term credit requires more papers to be produced by the farmer and more signatures of officials to be obtained. Long-term credit is even more of a theoretical facility for small farmers as only 19 percent had ever heard of the facility, 16 percent of how to go about obtaining the loan (the method being in fact identical to obtaining medium-term credit), and only one person out of sixty-three interviewed secured it.

The reasons for the poor performance of the SFDA extension services arise mainly through the vertical economic relationships in which the small farmer finds himself. First of all the small farmer, when attempting to secure a loan, takes quite serious risks. There are considerable opportunity costs arising from adopting changes in the rotation of crops or the deployment of labour resources (Hoskins, 1974 and 1975). Other risks, however, begin even earlier since farmers are competing for scarce resources against larger operators. The attention of the VLW, *mukhiyaa* or cooperative secretary, for example, is essential since his advice on form-filling and collection of title deeds of their land, his help in ensuring the smooth passage of the loan, and the delivery of seed and fertilizer (in the case of short-term credit) which is frequently late because of the long chain of funding institutions can determine the success of an innovation (Clay, 1975). But the smaller farmer, unlike his larger competitor cannot offer those who dispense these supplies comparable favours such as private loans at low interest, connections with government at a higher level, free accommodation and so on. A further point which makes the small farmer's position even more difficult is that he is obliged to commit himself at the outset to several innovative ventures, and therefore is in a weak position in subsequent transactions. For example, he must pay a share in the cooperative society before he can apply for a loan from that source and he may have to pay a bribe to the secretary even then. He may have borrowed a large sum of money (the interest of which may be financially embarrassing) to sink a tubewell but cannot obtain the electricity connection, and without the water cannot grow the crops to repay the loan. Two results follow from this situation. First, any small farmer finds it difficult to successfully negotiate a loan, and, second, a sense of cynicism and frustration is engendered in many by reports about the difficulties involved in obtaining credit. Over half the farmers expressed various versions of this attitude as a reason for non-application for short-term credit. This problem is exacerbated by the fact that many of the minor, local officials are themselves large farmers (Wood, 1975).

Wood (1975b) has shown—for a village close to the author's own study area—how membership of most of the village's organizations (Block Development Committee, School Committee, Caste Association, Co-operative Society, etc.) is drawn predominantly from the rich peasant population. Local officials are as much a part of the social structure (as opposed to the administrator's 'ideal' role of a non-partisan and neutral instrumentalist) as farmers themselves. Their office is used as a means of serving their own interests. Even non-elected local officers (the most prominent being the VLW) use their office to obtain facilities for those who patronize them. In any case, all those who are responsible for the diffusion of the package of information concerning the securing of loans for small farmers tend to depart from their official function of hierarchical diffusion and diffuse messages according to the biases determined by the social structure of which they are a part. While the Project Executive Officer (or his equivalent in non-IAAP Blocks, the Block Development Officer) is less part of the local power structure and is able to discharge his extension and administrative duties at the local level without prejudice, his level in the hierarchy determines that he has up to 25,000 farmers to cope with, if the verbal, face-to-face method of extension is adopted. Rapid tours of villages with brief talks to farmers at meetings (about which many farmers do not hear, as they are supposed to, from the *panchayat* servant/messenger) are usually the most he can manage. It is only a few PEO's who, by exceptional hardwork, really can affect the adoption of new inputs by farmers on a large scale. However, at lower levels of the hierarchy, the instrumentalist and neutral role of the extension agent becomes impossible to carry out and the hierarchical diffusion process, which is planned to reach target populations without social or geographical bias, is partially modified to reflect a form of neighbourhood diffusion which in turn reflects the social structure and associated resource distribution in which the local extension agent finds himself.

It must be mentioned also that hierarchical diffusion is not solely a phenomenon of the public sector. The diffusion of information and technical change often follows business networks, which are characteristically based on tours. As Harriss (1975) has shown tentatively in the South Indian case, towns trade overwhelmingly with themselves, and this tendency has increased markedly in recent years. Those business establishments which are the most recent, and serve the surrounding rural areas least, are those with the greatest starting capital and the largest absolute and per establishment turnover. The flow of goods from rural areas to middlemen with small business establishments and storage facilities and to larger dealers is a basic form of organization in most societies (wheat in the Punjab, Harriss 1973, or jute in Purnea District, Wood 1975b) and is the reverse image of manufactured goods which are distributed down the city hierarchy, with decreasing turnover and increasing price at each lower level of transactions. Such a structure, although not associated with a large urban–rural volume of trade, does imply strong linkages between large farmers and the city. *Zamindars* often invest their surpluses in urban property, in *godowns* (storage), cold stores, small businesses, shops etc., while many businessmen in the larger Indian cities invest in farms near the city. In both cases, technological change is transmitted by long-distance links between propertied and entrepreneurial persons. In effect, the type of diffusion is of a neighbour-to-neighbour type, except that 'neighbours' for the reception of the type of information implied here are situated in other urban areas or in rural areas considerable distances from the transmitter. Thus this type of diffusion is as much controlled by the social structure as neighbourhood diffusion except that in the former case 'neighbours' are not local, usually urban, and part of a widespread network underpinned by business interest, and in the latter, 'neighbours' are local, usually rural, and part of a 'neighbourhood' in the spatial sense.

These links can also be used in political confrontations between landlord and tenant. Wood (1975) mentions an interesting confrontation in Purnea District in which the local *Musahars* (previously 'untouchables') successfully organized a complete withdrawal of their labour from the absentee landlord. The landlord was able to break the strike by importing 'black-leg' labour from outside, using his wide-ranging network for mobilizing labour.

The Government of India has recently given prominence in the current Five Year Plan to the Integrated Area Development Programme. This programme is predicated on the belief that it can develop a form of urbanization which will provide planned facilities including finance, marketing, transport, extension services, communications, health, education and retail services. Such urbanization will constitute a powerful instrument for development and will operate due to 'the cohesive community which is formed by economic and social ties and physical restraints which relate a central place to a group of villages' (Pilot Research Project in Growth Centres 1973, page 6).

If the analysis outlined above proves to be broadly representative of what is going on in India, it is reasonable to expect that this policy will allow those with access to resources (that is those who already have them) to benefit further from technological change. Since most finance (particularly rural and small-scale urban finance), marketing, road transport and retail services are within the private sector and cannot be 'planned' within the context of democratic government in India, Harriss (1974) questions the validity of the concept of 'planned' intermediate urbanization. Intermediate urbanization will occur in any case partly as a result of the great increase in town-serving retail and manufacturing establishments. As such the business networks will all the more represent hierarchies of exploitation (each level of the hierarchy syphoning off surplus value from the level below). Associated with this hierarchy will occur a 'lateral' diffusion of information within levels of the hierarchy following existing trade patterns. Information on new economic opportunities, made possible by technological change within India, will tend therefore to reinforce the existing distribution of assets. The type of information which is associated with intermediate urbanization mostly concerns the technical details or sources of supply of such items as refrigeration plant, small steel presses, rudimentary plastic moulding equipment, electric motors for a multiplicity of uses in small industries, innovations in bottling, printing, spare-part manufacture for vehicles and agricultural machinery and so on. The vast increase in small industries such as these strongly implies a vigorous, well-connected network linking information, access to credit and supplies, operating at the city and medium-sized town level.

NEIGHBOURHOOD DIFFUSION

In a predominantly poor country such as India where three-quarters of the population is employed within the agricultural sector, the context in which a large proportion of individuals make decisions about innovations (in the widest sense, including the adoption of altered caste taboos, the new agricultural technology, family planning, etc.) is local and limited to a few sources of information. Where literacy and mass communications do not directly reach a large proportion of the population there is no nationally or regionally integrated pool of information accessible to the individual decision-maker. Much Indian empirical data supports this simple assertion. Mayfield's (1962) work on marriage distances, Roy and coworkers (1962), Singh (1965) and Shetty (1966) on agricultural innovations; the author's own work on family planning (1975) and Kivlin (1968) on health practices provide examples. It may be agreed that the differentiation of information stocks between individuals, families and groups of various definitions lead to a variety of cultures. Although the argument, if taken too far becomes ridiculously spatially deterministic, it may also be said that the more similar the information available to different persons becomes (the more freely available, the more identical in content), the more similar will be the decisions which they base upon it. The assertion could be elaborated upon to explain the relationship between metropolitan and regional cultures in more developed countries. Already interesting work has been done on the diffusion of dialects in East Anglia (Trudgill, 1972), which are directly related to broad verbal communication patterns which in turn are related to underlying economic forces determining journey to work and temporary and permanent migration from the countryside to towns and cities. In India there is one further important factor which serves to reinforce cultural heterogeneity, and this is the operation of caste observances. Whether caste is merely an aspect of superstructure to maintain unequal relations of production (Dumont, 1970), or is given a prime-causal role in the determination of these relations, caste serves to atomize society, and in its specific observances provide barriers or at least filters in the way information diffusion occurs. Specific instances in the neighbourhood diffusion of family planning and information on the provision of agricultural credit will be discussed later.

It follows that marked spatial variations of local culture due to limited communications over large (non-local) distances maintain themselves within the general Hindu culture, and are constantly changing throughout time. Accidents of history, local differences in ecology, physiography and so on are preserved because of the custom of endogamous marriage, short marriage links (between the homes of bride and bridegroom), commensality in certain defined cases and the very well-defined 'concretionary' web of relationships which connect the individual with his family, his lineage and his caste. For example, some of the remnants of Akbar's army retreating from a disastrous campaign in Bengal in which the army was decimated by cholera, settled in Bihar in a broad scatter following the general line of the army's retreat. Today their spatial distribution, distinctive title and 'caste' (although Muslim by religion) and kinship network is still preserved. Nearby in western Purnea District, many Santhal labourers (so-called 'tribal' persons from South Bihar) were brought to clear the forest after malaria eradication, and also to work on the Kosi River flood dykes. These people have continued to stay in the area, culturally distinct from other Santhals in the heartland of Santal Parganas to the south of the River Ganges, but have become dependent upon the landlords who imported them in the first place, and who acquired the cleared land often through illegal means. They have acquired some of the marriage customs of the Brahmins amongst whom they live, and their style of housing (although still distinctive in their new surroundings) has also altered as compared with that of their homeland. Slow spatial diffusion of items of information (which in extreme aggregate form might be defined loosely as 'culture'), the cultural heterogeneity which results from this and aspects of caste structure mentioned above have a most important effect upon class formation and the process of social change. First, the mode of production of the rural peasant defines him as locationally isolated, remote from regional or national organizations and the ideas they propagate.

In so far as millions of families live under economic conditions of existence that separate their mode of life, their interest, and their culture from those of other classes, and put them in hostile opposition to the latter, they form a class. In so far as there is merely a local interconnection amongst these smallholding peasants, and the identity of their interests begets no community, no national feeling, and no political organization, amongst them, they do not form a class (Marx in Shanin, 1971, page 230).

One of the most important elements in the articulation of class consciousness is the ability to transmit information through an organized network—a hierarchical network which is space-independent. The most efficient way to mobilize support for political action is by a hierarchical organization of party headquarters, regional offices, local cells, etc. (just as it is for government to launch a programme with a strong extension element). An earlier stage in the formation of this type of organization in North Bihar can be illustrated by an incident which occurred in Southwest Purnea District during September 1971. Fourteen Santhals were killed in a landlord-labourer confrontation at harvest-time. Within three days, nine thousand Santhals had congregated for a political rally in this remote, flooded part of Bihar, across the Ganges and over two hundred miles from the borders of the Santhal heartland. The message to congregate was diffused by a method identical to the one that was used at the Indian Mutiny in 1857—the hand-to-hand passing of a small *chapatti*, with accompanying message about where and when to meet. This event can be taken as representative of an early stage in the transition from occasional stimulus to institutionalized action in the form of a permanent political organization with party members, stated and widely disseminated objectives and so on. On the one hand the *chapatti* was recognized as a sign of extreme importance, and there was a sufficiently well-articulated informal message to activate nine thousand Santhals. On the other hand nothing was formalized at the rally (to the knowledge of the author) which would make diffusion of the call for political action involving complex messages easier next time. Perhaps it will require a few further discontinuous stimuli of this sort to bring these institutions (and their implied 'improved' method of information diffusion) into being. Elsewhere in India political expression of class formation and the organization of political articulation has been more developed, particularly in Bengal, Orissa and Kerala.

So far we have only suggested ways in which distance can affect the communications and organization of class consciousness, and have not discussed the major inequalities and their maintenance which is the main driving force of class consciousness. However, the way in which new economic opportunities are taken up as a result of diffusion processes tends to contribute to a widening income distribution, involving the commercialization of agriculture, the forcing of sharecroppers off land where no written lease exists and their reduction to the status of casual labourers. But, as has been shown in the Moroccan case (Blaikie, 1973), it is not so much the superior information networks of the rich farmers which leads to early adoption of the new inputs associated with the Green Revolution, but such farmers' superior access to ready funds, to loans (Frankel, 1971), and the scale economies of many of the inputs themselves (tractors, tillers, pumping sets, and most type of tubewells). Most of the literature on technological change in Indian farms hardly mentions the extension element and its diffusion amongst farmers (Johnson, 1970; Cleaver, 1973), except in the case of small farmers, where the access problem to both information and inputs has been shown to be a particularly intractable problem with given and largely unaltered property relations.

The lateral diffusion among small farmers about the range of new inputs and corresponding loans is a case in point. The SFDA's coordinating activities have been surveyed in a previous section, and it remains here to see how lateral diffusion takes the messages about loans, the means by which they are secured and the inputs for which they are used, from formal sources to the target population.

The selectivity of the SFDA's target population puts special limitations upon the diffusion of information because there are only a few 'small' farmers in each village. Some larger villages will have twenty (scattered in perhaps four or five *tolas* or small hamlets within the 'village'), while small villages may have only one or two. Because small farmers' neighbours may not be eligible for the SFDA's special range of loans, links in the diffusion process are missing. Although general technical information on farming may pass between farmers, the specific information on what kinds of loans exist for small farmers, the means of securing them, and the inputs associated with them is of interest only to small farmers themselves and as such is passed only between them. Of the 63 small farmers interviewed in nineteen villages in an area of about fifty square kilometres, only three had ever received information about short-term credit from other farmers of all categories, one of whom was a small farmer. Information about medium-term credit also was diffused to only three small farmers from all other farmers, of whom only one was himself a small farmer.

Apart from the issue of spatial isolation of possible interested persons, two other factors must be mentioned. The first is the nature of information about loans and new inputs. The information is quite complex and deals essentially with quantities at given points in time—in the case of a loan when it has to be applied for, the amount of loan, the proportion of cash and kind, the interest and period of the loan, the rate of fertilizer application, its timing, the timing and rate of other complementary inputs (such as water) and so on. Much of this information can be provided piecemeal through the extension agency; the farmer can be 'spoon-fed' with information and supplies together, a little at a time at the right time in the agricultural calendar. However, the successful diffusion of such information through the neighbourhood network must occur all at once—or at least in large connected messages—with the inevitable result that information from primary sources (extension agents, co-operative secretaries, *mukhiyaas* (village headmen) and Project Executive Officers, itself incomplete and haphazardly diffused (see above) is quickly forgotten by those who receive it and who have the opportunities to pass it on to others. A glance at the diagrams showing adoption outcomes for short and medium-term credit will show that 'other farmers' were cited as sources of information on the facility itself and the procedure of obtaining it in 4 percent of the interviewed small farmers in the case of short-term credit and 5 percent and 3 percent in the case of medium-term credit. No single interviewee received information on *both* the facility *and* the procedure from one farmer. Furthermore, preliminary technical advice on how to use chemical fertilizers which is given as part of the loan, was not received by the small farmers interviewed from any other farmer. We may conclude from this discussion that:

1. lateral diffusion is slow among small farmers, particularly in the case of SFDA coordinated activities because of the spatially fragmented nature of the target population;

2. the messages required for the successful adoption of credit are complementary and complex, involving interlocking pieces of information often of a quantitative type;

3. neighbourhood diffusion is particularly inefficient at disseminating this type of information package;

4. the net result of 1, 2, and 3 is that extension input has to be unusually dense, accurate and timely to fulfil the objectives of the SFDA; and

5. the agencies through which the SFDA work are unable to provide such extension input, to the extent that the stated objectives of the SFDA still remain largely a matter of words on paper.

Turning to the case of family planning, the situation in which diffusion occurs is quite different. An examination of the major determinants of these differences will then lead to an explanation of the different forms of diffusion and adoption patterns. A full account of family planning diffusion is reported elsewhere (Blaikie, 1975) and is therefore only briefly alluded to here. First, the target population consists of married couples from the age of 15–44 years, and therefore in almost every household, a target couple exists. Messages will therefore not be affected by spatial discontinuities of the networks through which they pass because of spatial isolation between nodes (would-be adopters). Second, family planning does not offer obvious private benefits as does short and medium-term credit. In fact for many poor families, family planning introduces serious risks into the fulfilment of many of their objectives, for example, income from a large number of sons working as labourers keeping the parents out of a destitute old age; sufficient sons who command dowries, and who will remain with parents in old age, unlike daughters who move to their husband's house after marriage; sons who will perform important funeral rites at the parents' cremation and so on, all compounded by the grave risk of existing children (sons being of particular importance) dying *after* a form of terminal family planning has been adopted (May and Heer, 1968). There is no competition for access to this innovation, and the poorer sections of the population do not therefore find themselves in the familiar contracturally inferior positions with landlords and local officials. Hence adoption can much more closely follow patterns of individual family situations (relating to parity, number of sons, man: land ratio, etc.) and to the quality and sources of information received by members of the family itself. SFDA credit-adopting patterns on the other hand are the result of additional administrative hurdles at each stage, and at which a cumulative proportion of aspiring adopters fall. Third, the adoption of family planning involves some very 'private' issues involving morality, relations and sexuality. The implications of the message therefore affects content (what can be talked about without embarrassment and what cannot), the composition of the dyad itself (which partly controls message content—a man will talk about family planning in a different way with his father than with a stranger in the bazaar), and therefore the speed and spatial spread of that message.

The spatially and socially selective way in which the family planning administration diffuses information has already been described. Given this distribution, how does the neighbourhood effect carry these messages to the target population? First, knowledge about family planning in some form was much wider among males with farms of similar size than about SFDA coordinated loans, in spite of the fact that extension outlets for agricultural information are far more numerous and are based in many more villages. (There are only about eight family planning workers or part-time workers who have other para-medical duties in most blocks in Purnea District who actually live and work in villages, while for farm loans there are theoretically sixty or seventy.) Second, not only is the network of information about family planning more connected than that for SFDA loans, but each transmitting node sends messages to more people. The estimated branching ratio of all family planning channels is 1:4.7 while for SFDA-related information it is about 1:0.04 (that is to say official extension is overwhelmingly more important than informal diffusion). In other words each person in the samples interviewed talked to about five other persons about family planning, while small farmers hardly even mention SFDA-related information to any other farmer, whether 'small' or not. Even more vigorous than general information about family planning, unfavourable information (stories of infection after vasectomy, of the death of parents, of the whole idea being against the wishes of God) branches at a rate of 1:10. Complaint, scandal and rumour move with astonishing speed throughout the country (one of the major reasons being that this type of information requires a much less strict selection process on the part of the teller about the receiver of the information, a point which will be expanded upon later). In fact rumours in general move with a similar speed. The author's own experience of the diffusion of knowledge of the death of Lal Bahadur Shastri, which was transmitted to the remotest village within hours, even to villages without a radio, is indicative here. Similarly the death of a man in the town of Forbesganj, the major town in the study area of the family planning survey, rumoured to be the result of a vasectomy operation carried out a week previously, was reported in a local Hindi newspaper. Although the readership of this paper was confined almost entirely to the town, within three days of the report in the newspaper, half of the interviewees in the survey which the author was carrying out at the time, had heard of the incident.

Secondly, information about family planning is diffused through two largely separate networks—one for husbands and one for wives. Interspouse communication in this backward part of Bihar (in terms of education for women, employment outside the home and other factors encouraging a more egalitarian relationship between husband and wife) was found to be very rare (supported by Poffenberger (1968) in a part of Gujarat State). Furthermore, since extension staff themselves have to work through two unisexual networks

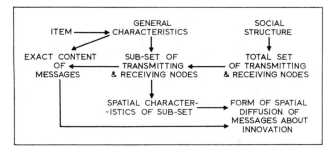

Figure 2.6 Schema describing factors influencing innovation diffusion

(male extension educators, and other para-medical staff such as sanitary inspector and antimalaria staff talking to husbands, while nurses and midwives talk to wives), the formation of two networks is assured. However the networks are unlike in most respects. The husband's network has relatively long links (in spatial terms) reflecting the husband's greater mobility; a much higher branching ratio; a moderate quantity of messages transmitted with low levels of message fidelity (that is considerable amounts of incorrect information and unsubstantiated rumour)—and by relatively heterophilous sources. The wife's network is highly spatially restricted, largely rural–rural, with a small quantity of messages but with high fidelity, and typically transmitted and received in dyads or small groups with the joint family or same caste (often whispered while waiting for water at the well or hand-pump).

To explain the rationality of the diffusion structures of these two different sets of innovations, it is necessary to put forward a simple schema which explicitly states the relationship between the item diffused, the quality and quantity of messages transmitted and the structure of the network itself. In Figure 2.6 the item diffused has certain characteristics which impinge on the social structure. Different items relate to different aspects of the same social structure, although in this case the aspects most relevant are the distribution of assets, political power and access to complementary agricultural inputs for SFDA-related innovations, and attitudes to dowries, discounted benefits arising from the birth of a son once he is able to earn money as a labourer, expectations of infant and child mortality, and belief in the importance of funeral rites performed by the son upon the parents' deaths for family planning-related innovations. The social structure in general determines the transmitting and receiving nodes for all sorts of communications—whether they are homophilous or heterophilous, in dyads or groups, whether information flows one way from one to many, or whether there is feedback, and the interconnectivity of the network within a group. The other major factor determining the *specific* transmitting and receiving nodes (with regard to a particular innovation) is the content of messages itself, although there is no clear-cut cause and effect. Nodes interact for many reasons, and the total set determined by the social structure (defined in the Indian rural situation by caste observances, perhaps nascent class consciousness, distance decay and the structure of marriage networks) determines the set within which messages about a specific innovation may pass. The word 'innovation' here can be used in the widest sense to mean new information about whatever subject. The subset of nodes which could transmit and receive messages about a specific innovation is defined by that innovation's general characteristics, while the precise content of the message is determined by the particular set of nodes at a particular point in time.

In this way a selection of specific messages about an item is transmitted from and to specific nodes, which themselves are a subset of all nodes which can interact according to probabilities determined by the social structure. In other words, a person meets many people from day to day, some of whom he/she might engage in conversation about a particular item. Even if that person does talk to another about this item, the way in which the conversation is handled, the messages which are transmitted and information which is known but not transmitted as a message is affected strongly by the relationship between the nodes within the larger context of the social structure. Thus messages start to travel through different networks. The nodes themselves, selected from the universe of population under study, have 'spatial characteristics' in that they have a spatial distribution of contacts (only a subset of which is relevant to the diffusion of a particular item), characteristic central places and so on. These characteristics determine the spatial form of the diffusion of messages.

Having outlined this schema and suggested some empirical evidence for it in the case of Purnea District, Bihar State, it remains to explore its limitations. It ceases to be a useful way of ordering the analysis of neighbourhood diffusion when the following conditions occur.

1. When mass media contribute significantly to the diffusion of the item. This is not so much because of the traditional caveat made by geographers that mass media is so much more distance-independent than face-to-face diffusion, but because they are far less subject to social structure—many persons can learn about an item whatever their socially determined position within information networks may be. Apart from this lower specificity of receiving nodes, the messages themselves are less subject to the relative position of transmitting and receiving nodes within the social structure. It is quite true that the content of messages transmitted by mass media is frequently controlled as a force in political manoeuvre, but the messages transmitted are not so carefully chosen for such small groups as in verbal transmission. The delicate ecology of message and node are both less strictly determined.

2. When social structure itself becomes less determined, in terms of the distribution of probabilities of contact between any one node and others. Because of the geographer's traditional interest in spatial biases in these probabilities (contact fields and movement fields), the related and more fundamental determinant of this distribution—the level of social differentiation—has been neglected. Simply, when social structure is less strictly determined, transmitters are less selective about those to whom they transmit and about the choice of messages transmitted.

Therefore it is in developed countries that the schema needs such a quantitative readjustment that it amounts to a qualitative change. In the case of India, it seems that for the most part these relationships have validity. However there exists an important feedback mechanism from improved communications (initially permitted because of a high level of literacy, and good public transport) to the articulation of class consciousness and political organization (for example, Kerala). In these cases messages can flow laterally through hierarchical organizations, and quickly at the local (neighbourhood) level. Large numbers of messages, amounting to complex arrangements such as points of view, reasoned argument and large scale reportage of events, can be transmitted and received intact.

CONCLUSION

India represents a wide spectrum of development of communication patterns. However the larger part of its rural inhabitants are linked by a primitive network of spatially restricted neighbourhood diffusion. Because of a low level of connectivity with high distance decay and high selectivity in most information-diffusing situations, networks are long and information has to pass through many nodes, rather than through a few at one level of a hierarchy to many at a 'lower level'. Messages therefore diffuse slowly, with a high rate of 'noise' and loss. This situation is as much a problem for extension agencies, as it is a major factor in the decay of class articulation and political organization. However, most of rural society in India is affected by the grass roots of metropolitan dominance, the lowest level of a hierarchical net, which distributes manufactured goods to the rural population and which purchases agricultural products for the metropolis. Such a network has all the advantages of rapid transmission of messages, as much because of its topological superiority as the length of its links. New technological innovations reach the nodes on this network rapidly, further deepening social differentiation, and political control can effectively be maintained at any scale (the few Santhals who tried too 'black' the fields of the *Zamindar*, up to the national political parties themselves). As metropolitan dominance becomes more marked, hierarchical organization of networks (along which flow information as well as influence and cash) become better organized such that the role of neighbourhood diffusion of all types becomes limited both in an absolute as well as a spatial sense.

NOTES

1. Purnea District is 4,200 sq. miles in extent, and its population is a little over 4 million (which is much a higher density than the all-Indian average). Only 8 percent of the population live in settlements over 20,000 so it is super-rural, even by Indian standards. The district is one of the poorest in Bihar as shown by most indicators of education, health, proportion of landless labourers, and amount of famine relief claimed. Bihar State itself is one of the poorest states of India. Muzaffarpur District is similar to Purnea

in many respects although it is less rural and better served by public utilities. However it is still below the all-Bihar average on most socio-economic indicators.

2. 18.00 rupees were equivalent to £1.00 sterling in July 1975. Hence 15 rupees would buy enough rice to feed a man for about four days.

REFERENCES

Blaikie, P. M. (1972) 'Implications of selective feedback in aspects of family planning research for policy-makers in India', *Population Studies*, **26**, 437–44.

Blaikie, P. M. (1973) 'The spatial structure of information networks and innovative behaviour in the Ziz Valley, Southern Morocco', *Geografiska Annale*, Series B, **2**, 1–41.

Blaikie, P. M. (1975) *Family planning in India: diffusion and policy* (London: Edward Arnold).

Clay, E. J. (1974) *Planners' preferences and local innovations in tubewell irrigation technology in north-east India* (Discussion Paper No. 40, Institute of Development Studies, at University of Sussex).

Clay, E. J. (1975 forthcoming) 'Choice of techniques: a case study of tubewells irrigation', *The rural problem in North-east Bihar: analysis, Policy and planning in the Kosi area*, ed. L. Joy.

Cleaver, D. (1972) 'The contradictions of the Green Revolution', *Monthly Review*, **2**, 80–111.

Dumont, L. (1970) *Homo Hierarchicus: essai sur le système des castes* (Paris: Gallimard).

Ford Foundation (1973) 'Pilot projects for integrated area development', for the Department of Community Development, Ministry of Food and Agriculture, Government of India. Mimeo.

Frankel, F. R. (1971) *India's Green Revolution: economic gains and political costs* (Princeton, U. S. A.: Princeton University Center of International Studies).

Harriss, B. (1974) 'The role of Punjab wheat markets as growth centres', *Geographical Journal*, **140**, 52–71.

Harriss, B. (1975) 'Rural–urban economic transactions: a case study from India and Sri Lanka', *Agriculture in the peasant sector of Sri Lanka*, Ed. The Ceylon Studies Seminar Committee, Sri Lanka.

Hoskins, M. (1974) 'The Green Revolution and cropping intensity', *Institute of Development Studies Bulletin*, **15**, 43–51.

Hoskins, M. (1975 forthcoming) 'Land holding and development in the Kosi area', *The rural problem in Northeast Bihar: analysis, policy and planning in the Kosi area*, Ed. L. Joy.

Johnson, E. A. J. (1970) *The organization of space in developing countries* (Cambridge: Harvard University Press).

King, M. (1972) 'Medicine in red and blue', *Lancet*, **1**, 679–681.

Kivlin, J. E. (1968) *Correlates of family planning in eight Indian villages* (Research Report 18: Project on the diffusion of innovations in rural societies, Michigan State University and at National Institute of Community Development, Hyderabad, India).

Ladejinsky, W. (1969) 'The Green Revolution in Bihar, the Kosi area: a field trip', *Economic and Political Weekly*, **4**, 1.

May, D. A. and Heer, D. M. (1968) 'Son survivorship motivation and family size in India: a computer simulation', *Population Studies*, **22**, 199–210.

Mayfield, R. (1962) *The spatial structure of a selected inter-personal contact: a regional comparison of marriage distances in India* (Technical Report 6, Spatial Diffusion Study, Department of Geography, Northwestern University).

Moore, B. (1969), *The social origins of dictatorship and democracy: lord and peasant in the making of the modern world*, (Harmondsworth: Peregrine Books, Penguin).

Planning Commission, New Delhi (1966) *The implementation of land reforms* (New Delhi).

Planning Commission, New Delhi (1969) *The problems of small farmers in Kosi area* (New Delhi).
Poffenberger, T. (1968) *Husband-wife communication and motivational aspects of population control in an Indian village* (New Delhi: Social Research Division, Central Family Planning Institute), Mimeo.
Prasad, K. (1966) *The economics of a backward region in a backward economy: the case of Bihar* (Calcutta: Scientific Book Agency).
Roy, P., Fliegel, F., Kivlin, J. & Sen, L. (1968) *Patterns of agricultural diffusion in rural India* (Research Report 12, Project on the diffusion of innovations in rural societies, National Institute of Community Development, Hyderabad, India).
Shanin, T. (1971) *Peasants and peasant societies* (Harmondsworth: Modern Sociology Readings, Penguin).
Shetty, N. S. (1966) 'Inter-farm rates of technological diffusion in Indian agriculture' *Indian Journal of Agricultural Economics*, 21, 189–198.
Singh, K. P. (1965) *A study of communication networks in sequential adoption and key communicators* (Unpublished Ph. D. thesis, Indian Agricultural Research Institute Library, New Delhi).
Trudgill, P. (1972) *Geographical diffusion models and explanation in socio-linguistic dialect geography* (Department of Linguistic Science, University of Reading), Mimeo.
Wood, G. D. (1975 a forthcoming) 'Administration and rural development in Kosi', *The rural problem in Northeast Bihar: analysis, policy and planning in the Kosi area*, ed. L. Joy.
Wood, G. D. (1975b forthcoming) 'The process of differentiation among the peasantry in Desipur Village,' *The rural problem in Northeast Bihar analysis, Policy and planning in the Kosi area*, Ed. L. Joy.

Chapter 3

Opportunity Space, Migration and Economic Development: A Critical Assessment of Research on Migrant Characteristics and Their Impact on Rural and Urban Communities

Alan B. Simmons

In recent years there has been an 'explosion' of research on migration, especially rural–urban migration, in developing nations of the world. In fact, the literature is so voluminous that reviews of findings tend to be either highly selective regarding issues, or they tend to be done on a regional if not a national basis.[1] This provides us with an abundance of findings on social change and the role of population movements in it. In this respect things have changed markedly from the situation less than 10 years ago when Breese (1966) in reviewing the available findings was forced to conclude that 'very little is known about factors which impel residents of rural areas and villages to make their way into the large cities' (page 79) or about rural-urban 'migration patterns' in general.

The abundance of findings is, however, not uniform across topics. Research carried out in recent years has emphasized some themes while others have been scarcely studied at all. Thus, Breese's (1966, page 84) assertion that 'little is known about ... temporal changes in the characteristics of recent, as compared with past, migration to urban areas' would appear to be as true today as it was 10 years ago. Gaps and lack of integration in certain areas of the existing literature make it difficult to draw out generalizations which would be of assistance to planners in developing countries.

The present paper seeks to provide a critical review of some strengths and weaknesses in the existing literature on internal migration in developing countries. Particular attention will be given to identifying topics, questions and hypotheses which need to be researched more thoroughly. While the review focuses on a limited number of issues, it has no particular geographical focus and reference will be made to illustrative studies carried out in various regions of the world, including Africa, Asia and Latin America.

The paper is organized into three sections. The first section outlines strengths and weaknesses in existing research, and the reasons why certain research themes have been studied more closely than others. The principal argument in this section is that many descriptive studies of migrant motives, characteristics and adaption have been carried out, yet relatively few studies have been carried out which link these variables to temporal change and regional variation in social–economic development. In addition, existing studies have paid relatively little attention to the developmental consequences of migration. The second section examines in greater detail hypotheses for future investigation on temporal and regional variations in age, sex, skill level and origin characteristics of migrants. The third and final section outlines some issues and hypotheses on the impact of selective migration on development in both sending and receiving communities.

The topics dealt with in this paper are based in part on an assessment of research priorities for development planners and policy makers. The first question for such individuals is whether or not existing migration patterns either ameliorate or exacerbate problems of increasing production or distribution of goods and services. The answer to such a question will depend in good part on the volume and characteristics of the migrants themselves and on how their residential change influences social and economic welfare in sending and receiving communities. Only after answers to such questions are found are serious attempts to influence current migration trends likely to be undertaken.

STRENGTHS AND WEAKNESSES IN EXISTING RESEARCH

Migration has been a favourite research topic in developing countries. There are several reasons why this is so, and these reasons help explain the biased emphasis given to some topics rather than to others in the available research. Two reasons in particular stand out.

1. Public visibility. Population movements are highly visible in developing countries in part because of their magnitude and in part because of their impact on political sensitivities in urban areas. Two-thirds or more of the adults in many of the large, expanding cities of the developing world are in-migrants, and their high fertility (due in part to their young age structure) means that the cities are growing very rapidly through the combined effects of continued in-movement and natural growth. Davis (1962) suggests that if the current rates of growth continue to apply, Calcutta will have a population of 35 million by the year 2,000.

Urban 'problems' emerging from this movement and growth have been particularly visible to politicians and planners. These problems include: the cost of extending services to the ever growing suburbs, the overloading of transportation facilities, water shortages, circles of slum housing around the outer periphery of the city, and general administrative confusion among overlapping political authorities within the urban area. The fact that a high proportion of the 'problem' population in slum areas may be migrants with little education and of poor rural background has reinforced stereotypes that *all* migrants are of poor, farm background. In the ensuing public debate about whether this stereotype is correct (as we shall see below it is not), and what should be done about population movements and rapid urban growth, there has been a strong demand for information about the migrants: their skills, their age, sex and marital status characteristics, their motives for moving and their employment and housing circumstances. Often this demand for information has been in the context of pressure to establish legislation intended to redirect existing population movements.[2]

2. Research convenience. In contrast with many other features of social and economic change, migration and urbanization are relatively easily measured, and can often be assessed from existing sources (census data) or from surveys of readily identifiable and accessible migrant populations (for example, the residents of urban low income or squatter housing). This research convenience is especially helpful in developing nations where information on labour force change, reorganization of production, land tenure, and other features of social and economic structure are limited and relatively difficult to obtain. One can count migrants and calculate the percent of the population living in cities as tangible evidence that technological and socio-economic change are taking place, even when the details of these other changes are only partially known. One can also survey migrants to learn indirectly about social–economic circumstances and difficulties in their place of origin which forced them to move; the advantages which have accrued to them in their new residence which keep them from moving back; their housing, employment, political activities and general sense of well-being, as well as their age, sex, and education characteristics which may determine the impact of their move on their place of origin and on their destination. Thus, one can learn about social, economic and political changes taking place in the nation as a whole by studying 'key' migratory streams. Not surprisingly, geographers, anthropologists, sociologists, economists, political scientists and many others, have all been interested in analysing some aspect of migration related to their own specialty. This has contributed to the volume of research on migration, as well as to the conceptual diversity of the studies undertaken.

Principal research strategies

Two principal research strategies have been pursued to answer the many questions about migrants and their impact: analyses of existing census data on migration, and special surveys of migrants. Each has its own strengths and weaknesses.

Census analyses

Census data have proved to be an important source of descriptive data on the general magnitude and direction of migration flows between 'provinces', 'states' or other such political–administrative regions within nations. However, census data provide only limited information on the origin, characteristics and personal circumstances of the migrant. The way in which data are gathered in a census typically restricts the definition of 'migrant' to those who are born in a different region from the one where they were resident at the time of the census. While the age, sex and other such personal data on the migrant, are available, nothing is revealed about when an individual arrived, how many places he or she lived in beforehand or the form of his or her employment history. This definition of migrant excludes those who move within the same political–administrative region, as might be the case when a person moves from a small town to a nearby large city. And for migrants who move between regions, one does not know whether the migrant's origin or destination is urban or rural, since the regions analysed will typically include settlements of varying size.

Despite these limitations, several useful studies have been carried out on the broad interrelationships between migration and social–economic context variables, using census data to measure interregional migration and other available data to measure interregional differences along several social and economic variables. The most common studies of this kind use regression techniques to 'predict' migration rates between the regions. For example, De Voretz (1972) showed that migration rates between states in the Philippines could be predicted by the additive effects of:

1. income differences between the states;
2. future income expectations, as measured by income growth rates in each state, and
3. the cost of the move and availability of information on destination, measured by distance between states.

Other studies, such as Godfrey's (1973) study of urban Kenya follow the Harris and Todaro (1970) model and predict migration on the basis of the cost of moving and the probability of being unemployed, in addition to income differentials. Schultz (1971) predicted migration between several localities in Colombia by employing a measure of levels of schooling in addition to measures of distance and wage differentials. Overall, these models have tended to predict interregional migration rates rather well; Schultz (1971, page 161) for example, predicts 49 percent of the variance in migration rates for the localities he analyses.

Nearly always the results of such regression studies are interpreted within a causal framework, in which social–economic variation is presumed to be a cause of variation in migration patterns. While such an interpretation generally makes good sense, it may tend to overstate the causality involved. The data for both migration rates and socio-economic levels refer to one time period in most studies (usually an intercensual period) and both migration and the socio-economic variables in this period may be dependent on a two-way interaction between them over a longer time span. Thus, one may ask whether low educational levels and out-migration are correlated because people want to live in areas with better educational opportunity, or because the better-educated people have over a long period of time tended to leave these regions? Sahota (1968) does not take into account the possibility of such two-way interrelationships in his analysis of the Brazilian case, and simply assumes that migrants are leaving for educational opportunities. Schultz (1971) in contrast looks at age-specific migration rates to test the hypothesis that the desire to leave for educational purposes will affect school-age migrants primarily. His results provide only weak support for this hypothesis, and hence lead one to speculate that education may simply be a general indicator of social and economic circumstances which is in itself influenced by prior migration patterns. While the interrelationship between wage and unemployment levels on the one hand and migration levels on the other would seem to be subject to a more straightforward causal interpretation, even this matter is not entirely clear since, for example, both wage and employment levels may be higher in areas which are attracting migrants, in part due to the capital and scarce skills which earlier migrants to the same destinations brought with them.

Special migrant surveys

Focused surveys of migrants have one central strength in relation to census data; they provide much more information on the characteristics of migrants, including detail on matters such as their employment and residential history, their housing on arrival to the city and over time up to the survey, and the involvement of their friends and kin in providing initial contacts and a sense of community within the city.

Migration surveys are limited, however, in so far as they have tended to focus almost exclusively on rural–urban migrants, often of low status and resident in slum housing, living in the largest cities. Nearly all the major cities of the developing world appear to have had their migrants studied, and not surprisingly, recent general reviews of the literature and edited collections on migration tend to focus on the evidence concerning migrants in cities (Brigg, 1971; and Mangin, 1970). It is also significant that the most comprehensive cross-national comparative study of migrant characteristics and migration consequences currently available in the literature relies upon surveys with residents of squatter settlements in eight countries: Indonesia, Malaysia, Nigeria, Philippines, Peru, South Korea, Turkey and Venezuela (IDRC/ Intermet, 1973). While several studies employ representative samples of urban areas which allow comparisons between migrants and non-migrants at all social strata within urban areas (Elizaga, 1970, Caldwell, 1969), only a very few studies of migrants in the city also include comparative data on non-migrants and return-migrants in the rural place of origin. Even when data on residents in rural areas is presented it typically comes from only one 'representative' town or rural area (Balán, Browning and Jelin, 1973; Simmons and Cardona, 1972). There are of course some studies which look only at migrants and potential migrants within the rural areas themselves. For example, Conning (1972) studied rural out-migrants in Chile on the basis of information provided by friends and relatives remaining in the town studied, and Eames (1967) studied migrants within rural communities in India. While these studies fill an important gap, they too suffer because they have no information on migrants in urban places. Herrick's (1965) analysis of a nationwide employment and migration survey in Chile, Caldwell's (1969) study of Ghana and the REHOVOT (1971) study of Israeli rural–urban migration are relatively exceptional, since they both include migrants and non-migrants in rural and urban settings. Perhaps unique, is the study currently being carried out in Thailand in which the location of migrants and non-migrants will be followed through time on a nationwide basis (Prachuabmoh and Knodel, 1974).

Available migration surveys are further limited because generally little effort has been made to link the survey findings the historical context of social-economic change in the region. Thus, most surveys tend to provide only static descriptions of the migrant population interviewed. Little attention is given to social changes which may have influenced the characteristics of the migrants interviewed, primarily because this would require a major independent data-gathering effort on the nature of such social–economic changes. Two exceptions, in which a direct attempt was made to relate survey findings to social and economic context variables, are illuminating in this regard.

First, Conning's (1972) study in rural Chile concerned the relationship between 'social differentiation', educational opportunities and other measures of community development, on the one hand, and patterns of out-migration, on the other. Census and other existing data was only partially adequate to assess levels of development in the community studied, hence Conning developed his own measures and scales through observation and interviews with key respondents in each village. Moreover, gathering data on the migrants themselves was complicated by the fact that they had already left for various destinations throughout the country at the time of his investigation. As a result, information on out-migration could only be obtained retrospectively from family members and relatives who remained in the communities. A rather complex and thorough set of questions had to be devised for residents in these communities to ensure that every out-migrant was accounted for. And, in coding the data, elaborate checks had to be instituted to make sure that out-migrants were not double-counted. While the data-gathering process was arduous, the information provided allowed Conning to make an interesting analysis of the links between general community development variables, educational opportunity, and the out-migration of young people to other villages or nearby cities. Of necessity the complex method of gathering data could be applied only to a limited number of villages, hence the findings are difficult to generalize.

Second, Browning and Feindt (1969) analysed data on rural–urban migration to Monterrey, Mexico in relation to the change in composition of the population in areas of out-migration. In this study careful attention was given to a cohort analysis of migration patterns which distinguished migrants by historical period of arrival. More recent migrants were found to have relatively low levels of schooling in comparison with the earlier 'pioneer' migrants. An analysis of census data on education in the surrounding rural areas showed that the early migrants came from more developed rural regions with higher educational levels in general, and the more recent ones were coming from the less developed rural regions with generally lower educational levels.

These two studies reveal how much additional effort is required to link migration survey data to the historical contexts of social–economic change. Not surprisingly, few studies of this kind have been carried out to date although future research along this line is obviously a priority if we hope to better understand the role of migration in the development process.

Needed research

As a result of the narrow urban focus and the lack of attention given to social–economic contexts in most migrant surveys, we know less than we ought to about both the determinants and consequences of migration.

Determinants

Most migration surveys are limited to analysing the 'reasons' which migrants give for their movement, to assess the factors which led them to move. As a number of reviews have indicated, adult male migrants typically focus on 'employment' and 'economic' reasons (Gilbert, 1974, page 115). Despite the dangers of *post hoc* rationalization being involved in such statements, and the fact that the urban samples interviewed generally exclude earlier migrants who did not stay in the city, the evidence is compelling because of the consistency across studies and because it agrees with the findings reported from regression studies using regional aggregate data on employment and wages as good predictors of migration rates.

However, we are still left with some important questions concerning the determinants of migration. Rural surveys in areas of out-migration would likely indicate many people who have difficulty finding work, and many others, whether working or not, who could improve their economic circumstances by moving elsewhere within the country. Yet only some of these people will move. Are those who leave different in terms of initiative, capital savings, or family and institutional contacts? If so, we would have a much more detailed understanding of the determinants of their migration. Most importantly, information of this kind would tell us more about the social impact of their departure on the sending community.

Consequences

Migration surveys are limited to analysing the consequences of migration in terms of the changes in social and economic circumstances among the migrants themselves. Thus, it is known that migrants experience social mobility (Balán, Browning and Jelin, 1973), improve their housing over time (Turner, 1969) and are generally satisfied with urban circumstances (Leeds and Leeds, 1970). However, we do not know how non-migrants in the sending communities, of the same education and age, have fared over time. Nor do we know to what extent the levels of satisfaction among those who have stayed in the city reflect the fact that migrants who were less satisfied have left, and were not included in the survey. Most importantly, the surveys give no direct indication of the social and economic impact of the migration on sending and receiving communities as a whole. At best, such impacts can only be assessed hypothetically on the basis of information on the age, sex and educational distribution of the migrant population. Studies which seek to measure changes in social and economic circumstances within sending and receiving communities over time, in relation to the magnitude and characteristics of migrants, are badly needed.

Hypotheses developed in the remaining two sections of this paper explore in greater detail specific issues on the relationship between migration and socio–economic development which deserve attention in future research. The first section advances arguments on the relationship between the changing characteristics of migrants over time in response to developmental changes. The second section concerns needed research on how the characteristics of migrants may influence development in sending and receiving communities.

WHO ARE THE MIGRANTS?

One of the first questions which emerges in developing nations experiencing massive internal population movement is: who are the migrants? The ample evidence on this question throughout the world shows that migrant characteristics are not universally the same, nor are they always strikingly different from the characteristics of non-migrants. As we shall describe in detail later in this section, migrants are both male and female, young and old, skilled and unskilled, from urban as well as rural areas. In some cases the migrant may have relatively more-skilled occupations, in other cases they may have relatively less-skilled occupations. Women may predominate in some streams and men in others.

This diversity in patterns of migrant characteristics may be initially distressing to those who seek simple universality in their generalizations. However, the difficulty one has in discovering universal patterns in migrant characteristics is hardly surprising since migration takes place under diverse circumstances in different countries at different stages in development. In the process of economic change, the spatial distribution of opportunities will change within the nation. Many opportunities in urban settings will be restricted to people with select qualifications, and of the people with the qualifications not all will be interested in pursuing the opportunities, even if they know about them. Individuals in a given setting may have systematically distorted information on the opportunities available elsewhere. Since the distribution of opportunities in the channels through which information about them is diffused may very with socio-economic organization, culture, and stage of economic development, it is perhaps more surprising that generalizations on migration characteristics can be made at all.

The discussion which follows is designed to show that while the characteristics of migrants are diverse, generalizations about them may be possible nevertheless in interpretations emphasizing the socio-organizational contexts (opportunity space) in which migration takes place. As we shall see there is in the available evidence at least the suggestion that certain general stages in economic development, urbanization and occupational specialization may be associated with particular patterns of age, sex, rural–urban origin or occupational skill in migrant streams. It must be emphasized that the arguments presented, while supported to some extent by available evidence, have not been for the most part carefully tested. They should be considered hypotheses for future investigation.

Age of the migrants

In various cultures and in less developed as well as more developed economic contexts, migrant streams are predominantly composed of young adults. The proportion of children and other people is subject to wider variation and may reflect the marital status or affluence of the young adult migrants. Young children, when they migrate, typically come with their parents. Cardona and Simmons (1975, page 16) noted a previously undocumented bulge after age 45 in the age distribution of migrants to Bogotá, Colombia. This may be a result of older parents coming to the city to join the children who have previously migrated.

The greater tendency for young adults to migrate is consistent with several fairly universal features of social organization in relationship to age. Young people are less likely to have a marriage partner, a tribe or a village position for which they are responsible, or to have established themselves in a trade or occupation. Consequently, they have fewer institutional ties to hold them and it may be expected socially that they go their own way and arrange their lives as they see fit. Moreover, not being encumbered with family obligations they are free to move alone and the cost of moving is lower.

The developmental process itself may create special circumstances which tend to encourage the out-migration of young people from the settled rural areas to cities or to 'frontier' regions in the nation. The following three hypotheses take different approaches to this possibility.

First, rapid population growth and large families may be one factor encouraging that out-migration of young people from settled rural areas. The reduction in infant mortality which has occurred and which continues to occur throughout the developing world has led to a marked increase in the number of living children per family. Cultural patterns in the rural areas in turn have developed over a long period of time under very different demographic circumstances, and are often unprepared to provide enough opportunity for all the children who now survive. As a consequence it seems possible that 'resource allocation' problems which increased family size creates for parents and other relations may be an important factor entering into family and individual migrant decision-making. It should be noted here that it is not necessary for there to be 'population pressure' on the land in order for a crisis to take place. Much will depend on social and economic organization. In a communal setting, such as that found in some African regions, there is obviously no problem for large families as such; there are only problems for large communities when their land resources become insufficient to meet everyone's needs. In areas of the world where land is privately owned, however, there may be low population pressure on the land but 'crises' within large families. This would seem especially likely if the large family is poor; yet it may occur if it is relatively wealthy and the members are forced to contemplate the division of family lands.

There is some limited historical evidence, particularly from Europe and from other developed nations to support the importance of the intermesh between demographic factors and social structure in forcing out-migration from rural areas. For example, Davis (1963) has argued that population growth associated with declining mortality and moderate to high 'traditional' levels of fertility in various European countries and in Japan was 'responded to' in different ways. In Ireland, the fertility was brought down through delayed marriage, while at the same time there was a heavy out-migration to cities in Britain and the United States. Japan, after the Second World War, when opportunities for migration out of the country declined and other changes in the opportunity structure for young people took place, experienced a very rapid drop in fertility, primarily through the widespread use of induced abortion. Following a similar hypothesis, Friedlander (1969) has argued that fertility fell earlier in Sweden than in Britain, in both rural and urban areas, because opportunities within Britain's cities and overseas colonies were expanding more rapidly and permitted 'strain' from rapid population growth to be relieved through internal and international migration. While the broad pattern of demographic and social events in these case studies does seem to support a general hypothesis that social and demographic interrelationships influence migration, it should be kept in mind that the studies do not specify precisely what constitutes 'strain' from rapid population growth nor how it becomes translated into a motivation either to migrate or to reduce fertility. If the hypothesis about 'too many people on the land' or 'not enough land to partition' were accepted as a measure of 'strain' then one might expect that rural people would be among the first both to reduce their fertility and to migrate. With regard to fertility at least it would appear that the urban middle classes were the first to accept contraception and intentionally lower their fertility in many countries (Banks, 1954).

Unfortunately, there has been very little systematic study of hypotheses linking population growth and out-migration in contemporary developing nations. Schultz (1971) notes that in Colombia areas with higher rates of natural increase seem to experience greater out-migration, even when income differences and distance between sending and receiving areas are held constant in a regression analysis. However, his results are not entirely convincing due to evidence of multicolinearity among the variables in the regression, which makes it difficult to come to any strong conclusion on the independent influence of any single predictor variable on the migration rates. Even if this problem did not exist, however, one would ideally want more direct evidence on how high natural growth and migration are linked. Are those individuals who leave rural areas with high rates of natural growth from the largest families and especially from those large families with limited resources? As we shall see below, there is evidence that rural–urban migrants are more likely to come from the wealthier rural families. This suggests that 'strain' from rapid population growth *per se* may not be a reason for their movement. Less is known about levels of out-migration and about the factors which influence these levels among the children from less advantaged rural families, since fewer of these migrants move to cities where the bulk of the research on migrant characteristics has been done. Such migrants may be more likely to move to other rural areas or small towns, and the extent to which 'strain' arising from rapid natural increase was a factor in their decision to migrate remains to be investigated.

Studies which examine the impact of rapid natural population growth under various developmental circumstances are required to test fully the impact of 'population strain' on out-migration. Operational measures of 'strain' need to be developed which show how population growth under specific resource constraints influences out-movement. Once this has been done, the problem can be put into a more balanced perspective. It will then be possible to ask how population growth and the scarcity of certain resources (such as capital and technology for more intensive and efficient farming) interrelate with migratory patterns.

A second hypothesis argues that development often has a direct impact on the 'control systems' (Mabogunje, 1970) which the family and village typically use to guide the activities of young adults in their society. Bantan (1957) in his study of tribal life in Freetown, Sierra Leone, noted that 'the young men stress that the foremost reason for coming to the city was that money was so easily obtainable there and that so many fine things could be bought with it' (page 57). However they further emphasized that in the city, they were free, whereas in their chiefdoms, they were subject to oppression and extortion (the Chief sided with the old and rich and prevented the young men with ambition from rising; they demanded more communal labour than the six days sanctioned by law, and if anyone protested he was victimized). 'Make I go Freetown; make I go free' was the attitude of many migrants.

The possibility that youths leave their home villages in part due to conflicts within the family or tribe, or more generally to escape parental and village authority, is worthy of further investigation. It should be kept in mind, however, that conflicts with authority may be only the visible manifestations of underlying stresses in culture, values and normative control systems, which are changing as a result of broad socio-economic and demographic change in the society as a whole. Primary attention should perhaps be given to these underlying stresses, since the quarrel which the migrant remembers as the reason for his departure may be a 'last straw' reason rather than an underlying motive. Thus, Bantan's (1957) respondents in fact may be migrating primarily for improved economic opportunities. While quarrels with the Chief may be symptomatic of a breakdown in traditional authority which makes it easier for them to leave, it may be an exaggeration to attribute the move itself to a desire to escape such conflicts. Or, perhaps even more correctly, the out-migration should be viewed in terms of a balance between the competing advantages and disadvantages of the home community in relation to other possible places of residence. Conflict in the home village would be just one element in such an approach.

Third, in the process of development educational opportunities as well as opportunities for mobility are frequently concentrated in the urban areas. The fact that young people migrate in greater numbers can be interpreted as the product in part of their parents' clear perception of the greater opportunities which exist outside the village. Conning (1972) has noted that in Chile, for example, the major factor governing the out-migration of young people from the villages was the desire to continue their schooling in larger urban centres. Chile has a good 'free' state school system, at least at the primary level, and rather high levels of literacy and schooling compared to most developing nations. In addition, it has a highly developed urban network (Herrick, 1965) such that initially the student need not move far, and may proceed in steps to larger centers. Whether aspiring adolescents who migrate to urban areas in other countries where literacy is lower and schools more costly also migrate in order to go on to further studies is not known, but would seem less likely.

One might conclude from the above points that one could reduce or stop the out-migration of young people from rural areas through a programme of intensive modernization, development and expansion of educational opportunities in the rural sector. The way in which Cuba was able to slow the process of urbanization by directing migrant streams from the urban to the rural areas in the middle sixties is evidence that this can be done, provided that it is co-ordinated by strong government control (Acosta and Hardoy, 1972). However, under other circumstances this may not take place. Bazán (1975) has noted that the sugar plantations on the Peruvian Coast which were nationalized and turned into cooperative farms under the current military government have undergone profound social changes, and now have better schooling, better health care and higher standards of living than previously. But, only the older generation are *socios* (members) in the cooperatives, and the young must leave in ever greater numbers to find employment.

It seems clear, therefore, that future investigations should be less concerned with simple generalizations about the fact that migrants tend to be young adults, and more concerned with the possibility that there may be substantial variations in the reasons why young people migrate, and correspondingly in the impact which their migration has on the developmental process. For example, in many more developed countries where educational opportunities are accessible but largely concentrated in large urban areas, youths may leave the rural area for study reasons and once trained they may not return home because there would be few opportunities for them there. In less-developed countries and at earlier stages in the development process, where educational opportunities are more restricted, youths may primarily leave the village for short-term labour opportunities and may often later return home with accumulated savings and relevant work skills. Thus, young adults may be the most prone to migrate in both settings, but the meaning of their movement for the long-term socio-economic development in their home communities is very different, depending on the level and/or historical period of socio-economic development. Hypotheses of this kind need to be examined directly through comparative studies.

Sex and marital status

The balance between men and women in migrant streams is somewhat variable. Jansen (1970) in reviewing a number of studies concludes that women predominate in some migratory streams and men in others, although there are more examples in developing countries of male migrants outnumbering female migrants.

Bogue (1969, page 764) has advanced the following hypothesis to account for the variation in the sexual composition of migrant streams. An initial stage of development is characterized by a preponderance of male migrants. Economic uncertainties are greater at this stage, insofar as the social norms in many cultures tend to 'protect' women by restricting their contacts to kinship networks and community, men are encouraged to take the risks associated with movement. In this stage the men may leave home for short periods only, or they may send remittances back to their families. Ultimately, the migratory process becomes routine and institutionalized. At this time male migrants are more likely to settle permanently and to bring their wives and children with them. Hence, the number of female migrants may begin to equal or exceed the number of male migrants.

Available evidence on the sexual composition of migratory streams does not allow any direct test of Bogue's hypothesis, but some findings are at least consistent with it. In Africa, where rural-urban migration patterns are more recent and where short-term 'labour force' migration and the sending of remittances is still common, it seems that men outnumber women in migratory streams. For example, in Ghana, there are 107 men to every 100 women in urban areas, while in rural areas there are only 91 men for every 100 women (de Graft-Johnson, 1974, page 476). However, even in Africa this may be changing since in recent times the female share of rural–urban migrants in Ghana has gradually increased (Caldwell, 1969, page 369). In Asia the pattern is mixed, with greater evidence of change toward female dominance in some migratory streams. In India it appears that men still dominate in migratory streams since men outnumber women in the cities (Eames, 1967). Prior to 1960, men dominated in migratory inflows to Bangkok-Thornburi in Thailand, but between 1960 and 1970 women predominated (Cummings, 1974, page 7). Females are apparently more migratory than males in the Philippines (Kim, 1972). In Latin America, where urbanization levels are high and migratory streams well established, women seem to outnumber men (Gilbert, 1974, pages 113–114). In addition, there is information from various settings that migration is increasingly made up by the entire nuclear family (IDRC/Intermet, 1973, page 74). The urban migrant family may take on a more extended structure in order to accept migrating kin, including unattached women who might not otherwise have migrated (Bruner, 1970, pages 130–132).

Despite the general fit between some findings on the sexual composition of migrant streams and Bogue's hypothesis, it is apparent that the hypothesis is inadequate to deal with some of the evidence. Specifically, some evidence fits more closely to the hypothesis that the proportion of women in migratory streams depends on labour force opportunities for them much more than it depends on the 'risk' involved in their move. Consider the following arguments.

The fact that women follow more institutionalized and less risky migratory streams is often felt to be a reason for the generalization that women tend to move shorter distances than do men. Ravenstein (reported in Lee, 1966) was so convinced that this was the case in Europe, that he offered as one of his 'laws of migration' the assertion that females are more migratory than males in short distance moves, while males are more migratory than females in long distance moves. However, the generalization is hardly a law since one can find exceptions. The exceptions emphasize the fact that neither women nor men are inherently more likely to take 'risky' long distance moves, since everything depends upon the organization of social and economic opportunity. For example, in Peru women tend to predominate in long distance moves and men predominate in short distance moves (Alers and Applebaum, 1968, pages 28–29). This long distance migration of women in Peru is quite likely a function of the market for domestic servants. The large cities (particularly Lima) are relatively distant from the populated rural areas where domestic are recruited. Yet there were some 90,000 women in domestic service in Lima alone in the period 1966–1967 (Smith, 1973, page 193). These women represented close to 10 percent of all women in the economic ages, 15–60 years, and over 60 percent of all women who actually worked in Lima. Not surprisingly, women predominate among migrants to this city (Lowder, 1970) apparently in response to the great demand for their services. When one considers that Lima is an archetypal 'primate city' with approximately one quarter of the nation's population living in it, the profound impact that the opportunity structure for domestic servants has on migratory patterns for women must be evident.

In other Latin American cities a similar situation is found. An exceptional case has been noted among migrants between 15 and 24 years to Santiago, Chile, where there were only 64 men to every 100 women (Herrick, 1965). This exaggerated pattern cannot be explained merely by the fact that migration has become stable and less risky for women. Even in the Asian examples where women predominate among migrants it appears that labour force factors are central explanatory variables. Kim (1972) argues that women predominate in migratory streams in the Philippines due to the fact that the country is more westernized and more modernized than other neighbouring countries and Philippine women are perhaps more likely to pursue careers independently in places other than those of their birth' (page 16). Cummings (1974, page 7) also suggests that changing opportunities for women in domestic and other service jobs may account for the increasing proportion of females migrating to Bangkok-Thronburi in Thailand.

It seems clear from the materials reviewed here that previous research has been focused primarily on simple descriptions on the extent to which males or females predominate in migrant streams. In contrast, the more fundamental questions of why various patterns should exist and what the consequences of these patterns are for sending and receiving communities, have scarcely been asked. As a result, the hypothesis argued here that the sexual patterns of migrants are closely bound up with the role and status of women in different societies as they change over time in the developmental process, requires further investigation. From a policy-oriented research perspective the key issues to which future research should be directed include the role and status constraints on women in different cultural and developmental settings, the ways these are changing, and their implications for the patterns of female labour force participation and female migration. As women become less dependent on and restricted to family roles, will they migrate more? It is also important to know whether this migration will have any impact on their labour force participation?

Skill levels

There is a high proportion of migrants in the socially visible 'marginal housing' of large cities in developing nations and this often leads to the mistaken impression that all migrants are in the lower strata. In fact, there may also be equally high proportions of less visible migrants in other social strata, including the élite. This is the case because migrants to large cities tend to be drawn disproportionately from other (smaller) urban centres and from higher status families in towns and villages. As a consequence, the migrants tend to have higher levels of education and more skilled occupations than do those who remain behind. There is widespread variation in this generalization from one region of the world to another: in Latin America the generalization holds rather well in most instances (Cardona and Simmons, 1974), yet there are some cases where very recent migrants seem to be among the least educated and skilled in their places of origin (Browning and Feindt, 1969). While the better educated are more likely to move to the cities in India, it is nevertheless true that there are important mechanisms for recruiting lower caste workers into urban sweatshops (Bogue and Zacharia, 1962). In Africa, individuals in the minority group with higher levels of schooling are also more likely to move to the urban areas (Caldwell, 1969, page 69); yet it is nevertheless the case that most migrants have low levels of schooling because schooling levels are low in the sending communities. Thus, while rural–urban migrants tend to be more educated and more occupationally skilled than those who remain behind, this is always just a matter of tendency, since migrants nearly always reflect the skill levels in their place of origin.

Balán (1969) has suggested that the selectivity of migrants and the impact of migrant characteristics on the city are both determined by level of economic development. In an early stage of development urban centres are small and tend to serve administrative, rather than productive and distributive, roles in the economy. As a result opportunities for urban employment are very restricted. Any rural–urban movement at this stage, according to Balán's model, is more likely to involve the sons and daughters of the rural land-owning élite who come to the city for education and perhaps to take on a professional or administrative job. There will be a few opportunities for people of lower social rank to fill in selected positions in service occupations. Later, more urban employment opportunities emerge but information about them spreads slowly to the rural areas, transportation is difficult, and the villagers in the rural sector do not have urban occupational skills. Children from non-élite families begin to move increasingly into the cities for education and for employment, but at this stage migrants are still rather selective of families with at least some wealth (owners of small and medium size farm-owners, store-owners in the villages, etc.). These are also people with higher educational aspirations and greater contact, through commerce, with urban life. Eventually, services and construction in the urban areas expand, communication between the villages and the city increases and population growth and/or technification in the rural areas combine to increase the relative attractiveness of the city for peasants. Rural–urban migrants at this latter stage come with increasingly less education than those who migrated at a previous stage.

Consistent with Balán's argument is the hypothesis that the distribution of skills which migrants have and the size and occupational complexity of their places of origin may be interrelated. In those less-developed nations where all villages are small and dependent on agriculture, there will be little occupational differentiation. Migrants to large cities will necessarily be of rural rather than small town origin, and if the number moving is substantial they will tend to be occupationally undifferentiated from non-migrants in their home town.

But in those less-developed nations with more highly integrated networks of towns and regional cities, a higher proportion of migrants to large cities may be urban in origin and may have urban occupational skills. Herrick (1965), for example, has noted this to be the case in Chile. Where educational systems and specialized occupations have penetrated rural areas the occupational characteristics of migrants can very more widely and have a more dramatic impact on both sending and receiving areas.

The research in support of this 'developmental' approach to understanding migrant characteristics is rather limited and is largely confined to more occupationally differentiated developing countries, particularly those in Latin America. In reviewing selected historical data on characteristics of migrants in six Latin American nations at distinct levels of urbanization and socio-economic development (ranging from Guatemala to Argentina) Balán (1969) found some support for his general hypothesis. Browning and Feindt (1969) show that migrants to Monterrey over the past several decades have shown marked changes in educational and occupational characteristics such that recent migrants are much less skilled (in absolute terms) than those who came earlier. The early 'pioneers' had to overcome barriers to arrrive in the city and by implication were able to move in part because they had the resources to overcome these barriers. The current 'mass migration' of the relatively disadvantaged people in rural areas, whose movement is facilitated by the fact that migration is now an institutionalized process, in which most potential migrants have friends and relatives in the city who can assist in the transition to life in the city.

Extending upon Balán's argument, Browning and Feindt (1969) have further argued that at this latter stage only those with financial resources and technical skills may be able to survive in the rural area, hence selectivity may actually be 'negative' with the least capitalized and most poorly educated dominant in rural–urban streams. However, this is only an inference from the changing composition of rural–urban migrants from areas surrounding Monterrey to the city itself. No measure of the educational and occupational background of all out-migrants from this rural zone is available, hence one does not know if the better educated, more occupationally skilled out-migrants from this predominantly rural zone are now destined for even larger urban centres, such as Mexico City itself, where the opportunity structure may be even more open to skilled migrants. In addition, there is no direct evidence in the study on the changing structure of income opportunities of unskilled and semi-skilled works (as might take place through farm mechanization) in the sending region which would lend support to the hypothesis that the least skilled are being, in a sense, 'forced' to leave.

The Balán approach may be most appropriate for countries in which all urban growth is concentrated into 'primate' cities. It may be less appropriate for nations in which there is an urban network of towns and intermediate cities. To take a Latin American example, Simmons and Cardona (1972) have noted that the occupational and educational levels of migrants to Bogota, Colombia, have not changed over the past forty years. Recent migrants still have rather high levels of education and occupation relative to non-migrants in the major origin areas. As a result few of the small, near subsistence-level, farmers from nearby areas move directly into Bogotá: whereas 68 percent of the highland population surrounding Bogotá lives on farms and in isolated rural hamlets, only 20 percent of the migrants to Bogotá come from these farms and hamlets. Most of the migrants to Bogotá, then, are from the towns, villages and small cities in the region. However, for the most part these towns, villages and small cities are continuing to expand although the growth is generally slower. McGreevey (1968) has hypothesized that this growth may be due to migration of the farm population into the towns and regional cities. He further hypothesized that some but not all the emerging opportunities in these towns which attract the rural population may be created by the departure of former residents of these towns, as they in turn move to larger cities. Unfortunately, little is known about the characteristics of migrants to small towns and intermediate cities in Colombia (or elsewhere). But, since the in-migrants to the towns predominately originate in rural areas where occupational skills and educational levels are lowest, one may infer that they are less skilled and educated than those migrants going to large cities (Adams, 1969). If this is the case, and these towns and cities continue to grow by absorbing those migrants leaving the farm sector, then Bogotá as a principal urban centre may never experience the latter stage suggested in the developmental model (and possibly supported in part by the case of Monterrey), where less skilled migrants from farms and isolated hamlets flood to the metropolis.

It should also be noted that a developmental model of migrant skill appropriate for Latin America, may be inappropriate for other regions of the world. Latin America is characterized historically by extreme differences in wealth, concentration of land in the hands of the minority, a small but important commercial sector, and a highly dependent peasantry. West Africa, to take a contrasting case, has been characterized historically by communal lands and less dependence on monetary and commercial systems. In some parts of West Africa, rural population growth and technological change (not to mention ecological disasters) are creating the impetus for major migrations long before the urban centers are prepared to absorb them. As a consequence, much African migration is rural–rural, rather than rural–urban. Moreover, actual patterns of rural–urban migration reveal far less status selectivity than migration patterns in Latin America. In the case of mining or plantation towns, in-migrants typically go for a short period of time and may subsequently return to the villages, while at the same time their place is taken in the city by a new migrant. These processes in West Africa all suggest that current urbanization in this region is not comparable with the kind of urbanization which is now taking place in Latin America, because the opportunity structures are different. An interesting question is whether West African patterns will begin to approximate Latin American patterns if ongoing economic and social changes lead to increasing private and corporate ownership of rural lands, greater occupational stratification and dependence on wage-employment in the farm sector, and more larger, more diversified urban networks. There seems to be some evidence that the frequent back-and-forth movement of workers in West Africa is declining. Female and family migration may be more prevalent now than previously and unemployed men in the city seem to stay there rather than go back to the rural area (Gugler, 1974).

Other characteristics

Migrants may be distinctive in a number of ways other than sex, age or skill level. Some of these characteristics are likely to reflect the socio-economic and cultural contexts in which development is taking place. For example, it has been widely noted that distance tends to play an important part in the migratory process, with most migrants to a given area coming from the nearby settled regions. As previously indicated, several regression models of inter-regional migration find distance to be an important predictor of migrant flows. While there is little direct evidence on this matter for the developing countries, the general historical pattern in the developed regions of the world suggests that distance becomes less important over time as transportation facilities are improved, real removal costs lowered and information spreads through the country as a result of expanding road and communication systems (Lee, 1966).

Certain ethnic or regional groups are more likely to move than others, and this may reflect occupational specialization in these groups or other cultural characteristics which facilitate or inhibit their movement. Textor (1956) has noted that the pedicab operators in Bangkok tend to come from certain villages in northeast Thailand, possibly because recruitment into this occupation is through personal contact in the home villages, and individuals in these villages' already know a great deal about the occupation and what to expect in Bangkok through contact with relatives already resident there. A contrary example is provided by Lowder (1970) who wonders why there has been so little out-migration from the Puno region of highland Peru in the light of the fact that the region is densely populated and has suffered heavily from droughts and famines over the years. A likely hypothesis concerns the fact that the Indian population of the region remains isolated from Lima and other urban growth centres in the country due to its distinctive language and culture (Gilbert, 1974, page 109). How and when the cultural 'isolation' of groups such as the Puno Indians will begin to break down may depend heavily on future patterns of economic and cultural penetration into the region from outside. A general hypothesis for future investigation is that migrant streams to the city in multi-ethnic societies will become increasingly heterogeneous over time, as communication and information networks expand.

THE CONSEQUENCES OF MIGRATION

There are several reference points for evaluating the impact of migration.

First of all, the individual migrant may undergo economic, social and political changes as a result of moving to a new location. Secondly, the community of origin may suffer or benefit as a result of the process of selective migration. Thirdly, the receiving community may suffer or benefit as a result of gaining migrants with high skill or education levels. Finally, the economy of the nation as a whole may benefit or suffer from changes in population distribution. The first three possibilities will be considered in the following section, but the last possibility goes beyond the scope of the present paper and is not discussed.

Just as the volume, pattern, and characteristics of migrants may very with the level of socio-economic development, the effect of migration may also vary with such circumstances. To reduce the number of alternative patterns to be considered, we will focus on one important and, in some regions, increasingly common pattern of migration. Our focus will be on an intermediate stage of urbanization in which migration from rural areas and small villages to cities and larger urban areas is numerically the predominant pattern. Such an intermediate stage may be preceded by a stage in which rural–rural migration is predominant (as it is still in parts of Africa) and it may be followed by a stage in which urban–urban migration is predominant (as it may be becoming in parts of Latin America). This stage is currently important in Latin America, dominant in many nations of Southeast Asia, and increasingly evident in Africa.

Impact on the migrant

More direct information is available on the effects of rural–urban migration on the migrant than on the consequences for urban residents or for non-migrant villagers. The existing findings strongly support the conclusion that rural–urban migrants are, on the whole, pleased with their move. The evidence is very strong for countries in Latin America and Asia that migrants move from areas with lower wages to areas with higher wages. Not surprisingly, the migrants themselves typically indicate that they move for reasons of economic improvement or economic necessity. In Africa, where the existence of traditional markets makes the calculation of wage differentials difficult, the picture is not so clear (Gugler, 1974, pages 7–9) and it may be that the availability of certain social services in the urban area, such as schools, medical facilities, piped water, etc., may be crucial factors in attracting the migrants above and beyond wage differentials. Perhaps because they are relatively skilled and in productive ages, migrants typically find jobs quickly in the city (IDRC/Intermet, 1973). At least their levels of unemployment seem no worse than those of the urban born. In fact, better educated migrants, many of whom arrive young to continue their schooling, may have substantially lower unemployment rates and higher occupational mobility than the urban born (Balán, Browning and Jelin, 1973; Simmons and Cardona, 1972).

It is always possible, of course, that the economic adaptation and increasing security which seem to accompany longer residence in the city are merely a consequence of inadequate research frameworks. Most of the results on this topic are based on 'one-shot' surveys, in which current employment status is related to the length of time since the migrant arrived. This approach fails to take into account the possibilities that either over time the less successful migrants may have returned, leaving only the more successful, or that the early migrants arrived in easier times or with superior work skills. One study which examined these possibilities found that:

1. return migration to the rural area was sizeable but included a predominance of the most successful rather than the least successful migrants, and

2. that recent as well as early migrants obtained their first jobs in the city with equal facility (Simmons and Cardona, 1972). However, given the 'case study' nature of the rural sample, and the small size of the urban study on which the period of arrival data for migrants was based, confirmation of these findings is required for Bogotá and for other cities before any generalizations are warranted.

Even poor migrants may find that the city allows them to increase their security (Dorner, 1964, page 248). When poor migrants, for example, turn to squatting to solve their housing problems, they may find that, in the context of urban growth and new migrants arriving to the city, they have a central and relatively secure residential location. They can rent rooms in their homes as they expand them. The house itself can be used for petty commerce. Squatters are generally found to be fairly satisfied with their communities (Leeds and Leeds, 1970; Flinn, 1968).

To the outsider the rough, crowded, small houses in squatter settlements, the open sewers and unpaved streets of many marginal urban communities in developing nations may seem wholly unsatisfactory. Yet the IDRC/Intermet (1973) survey of several low income communities in large cities of Africa, Asia and Latin America, found that electricity (sometimes 'stolen' from passing lines), water (piped at least to the neighbourhood), refuse collection, and health care were available at rather high levels, and this undoubtedly contributes to the benefits which migrants obtain from the city.

The above evidence should not be taken as an indication that everything is easy for migrants. Those who arrive without resources continue to be the poorest in the city, getting by on marginal jobs and suffering from very high levels of unemployment. Unemployment and underemployment among this group may be no higher than among the poor urban born, but it is still intolerably high. Gugler (1974, page 2) points out that surveys in nine African nations report unemployment rates ranging up to 20 percent of the labour force and of those who are employed many have marginal jobs, such as boot blacking or hawking in the streets. In some cases unemployment seems much higher in urban areas than rural areas, although the unemployment statistics themselves are highly suspect. K. T. de Graft-Johnson (1974) reports a correlation coefficient of 0.61 between rates of unemployment and urban size in Ghana. If true, a crucial question is why migration continues in the face of widespread urban unemployment, and why many of the urban unemployed do not return to the rural area. It would appear possible that the vast majority of the unemployed in urban as well as rural areas do *not* move even in the face of prolonged and severe hardship 'partly because they believe economic conditions are no better elsewhere' (Morrison, 1973, page 23). Of course, many of the unemployed in urban settings are the younger, less-established workers who had left the rural areas in search of work just as they were entering the labour force. They never contributed to rural unemployment rates, and it is likely that they would simply transfer their unemployment to the rural areas were they to return. Given the choice of where to be unemployed, many migrants may prefer the urban sector, both for the services available and for the higher wages they will get if they do find work. This would be encouraged if there were relatives in the city who could provide shelter and support during the period of underemployment or unemployment. It should be emphasized that so little is known about underemployment, productivity, and wages in rural Africa (as well as many rural areas in Asia and Latin America) that it is often impossible to say whether the underemployed migrant is better or worse off in the city than he would be in the rural area. This is a topic on which more research is badly needed.

One consequence of migration to the city concerns the educational value of simply living in the city and being exposed to a wider range of information and images of the world. Simmons (1970) discovered that less-educated rural migrants living in Bogotá showed considerable gains in general information over their first decade of residence in the city. In fact, each three years of urban residence over this decade increased their scores on vocabulary and general knowledge tests by an amount somewhat greater than the increase that would be expected had they gone to school for an additional year. In this way, rural–urban migration contributes to the personal skills of the migrant, and these skills may have repercussions for employment and occupational mobility in the country as a whole. This is another hypothesis which requires further elaboration and investigation.

Impact on the urban destination

The consequences of migration on the urban areas in developing nations has been the subject of a great deal of speculation but only a small amount of direct research. While the age and labour force characteristics of migrants are relatively well known, the extent to which cities really benefit by receiving migrants with these characteristics is not clear. Those who migrate may be relatively young, and better educated than the average population in the sending community. Despite this selectivity in the sending community, the 'average' migrant may still frequently be less well educated than the native in the receiving urban community (Simmons and Cardona, 1972). Since rural–urban migrants are better educated than the average person in the sending communities and less well educated than the average person born in Bogotá, their change in residence has the anomalous effect of reducing the average level of schooling in *both* the sending communities and in the city.

Certainly the cities would seem to gain from having received a number of skilled people whose services are needed in, and whose training was at no cost to, the urban area. At the same time the cities often receive more migrants, especially migrants less skilled than the average urban worker, than they can provide employment for. While these people might not be productive in the rural area either, the burden of supporting them in large numbers may be heavier in the urban centres. Unfortunately, no system for balancing the gains in needed migrant workers against the losses in supporting underemployed migrants has been devised to assess the net benefit or loss to cities, nor any way of estimating how such gains and losses vary under different developmental circumstances.

The increase in urban population due to migration may provide certain economies of scale in providing needed services. A few efforts have been made to determine the per capita costs for minimal services such as water, sewers, roads, housing and education in urban areas. For example, the World Bank (1972, page 4) estimated that per capita costs for installing such minimal services were in the range of U.S. $500.00 for a typical urban area in a developing country. Another study, using data for India, suggested that per capita service costs for industry decline with increases in population until the city is in the range of 130,000 to 300,000 after which the size of the city may again push up per capita costs (Stanford Research Institute, 1969). It should be emphasized however that the 'optimum size' of cities for achieving minimal per capita public service costs will vary substantially depending on a wide variety of assumptions and conditions: availability and space utilization of land, type of sewage system (water-bourne systems may be particularly expensive in arid nations), and so on. Studies in this area need to be improved and expanded considerably before any conclusions can be reached. The most that can be said at the present time is that migrants provide a mixture of costs and benefits to the city, and that per capita public service costs may possibly increase after a certain (unknown) population size is reached.[3]

Turning from economic to social factors, there has been a frequently voiced fear that less affluent migrants with employment problems might become violent criminals and political dissidents. There is a limited amount of information which provides some basis for this fear (Breese, 1966, page 97), but most of the available evidence does not support it. Migrants seem to be no more likely to engage in crime or radical politics than other citizens (Cornelius, 1969, 1970) and residents of many squatter settlements set up policing systems in their own communities (Abrams, 1966). On the basis of available evidence, then, there is no reason to believe that either political violence or crime is inherently higher among migrants than among other members of the community, although residential crises and employment difficulties may lead migrants to take certain forms of action such as squatting on land or vending without a licence, which may in fact contravene the law.

One of the few identifiable social problems to be associated with migration is the potential for ethnic conflict in multi-ethnic societies. For example,

Withington (1967) has documented the discontent among many native Sumatrans regarding the in-migration of Javanese who practise different forms of agriculture and occupy unused land. Other similar conflicts are widely known but undocumented in the research literature. These would include the quarrels in Kuala Lumpur between Malays arriving to take over positions in government and the economically dominant Chinese; the 'Civil War' on the island of Mindanao, Philippines, arising in part from the settlement of Christian Visayans in this traditionally Muslim area; and migration and ethnic interpenetration as one of the factors leading to the recent Nigerian Civil War.

It would seem fair to conclude that much of the concern about the possible negative consequences of large city growth has not been supported by research carried out to date. Government policy makers who have promoted laws and other measures in an attempt to limit population growth in large cities appear to have done so out of rather personal reactions to the appearance of 'crowding' in these cities, and from fears that mounting pollution, traffic congestion, and other urban ills will eventually destroy the quality of life in the city for everyone. These are very real concerns, however they are expressed largely in terms of emotional and psychological costs which are not easy to quantify. And, they are not easily related to the substantive economic costs and benefits to the city itself, which have yet to be investigated.

Impact on departure areas

Although out-migration generally benefits the migrants themselves, 'there is no satisfactory way to measure how their departure benefits or harms the remaining residents' (Morrison, 1973, page 22). This being the case the most that we can do in this review is try to evaluate several *possible* outcomes for out-migration communities.

Certain positive outcomes for the sending areas have been hypothesized. One possibility is that unemployment and underemployment may fall in the sending area as migrants leave. However, the impact of out-migration on unemployment may be small given the age and occupational characteristics of the migrants. As previously noted a high proportion of the migrants are in their late teens and early twenties, and may not be firmly established in the labour forces of their areas of origin. Their departure may not reduce the unemployment rate although it may keep it from rising. A related argument is that out-migration will lead to a decline in the supply of labour which, given a constant level of demand for labour, should lead to rising wages. Unfortunately, this does not seem to happen either because even after the migrants have left there still tends to be a surplus of rural workers. More efficient ways to raise rural wages would seem to be eliminating this surplus labour, perhaps through a combination of out-migration and land reform, and the introduction of labour-intensive technologies. At the present time too little is known about labour force/migration inter-relationships in rural areas, and they remain topics for future research.

Once they have left and found work elsewhere the migrants may contribute to the sending community in substantial ways. In Asia and Africa the practice of sending remittances to one's family in the rural area persists. While statistical information on this topic is not always soundly based, Caldwell (1969, page 168) has estimated that remittances in Ghana in 1963 may have been as much as £16 million, or 3 percent of national income. He further estimates that of the £5 million total income earned in Accra, as much as one tenth is sent out as remittances, primarily, of course, to rural areas. Almost 90 percent of a recent migrant sample in Nairobi, Kenya, claimed that they regularly sent some money home, and that the average sent amounted to approximately one fifth of their income (Johnson and Whitelaw, 1974, page 474). The impact of these remittances on rural areas may be substantial. Gaur and Nepal (1962) and Visaria (1972) found that recipients of remittances in rural India were able to open bank accounts, acquire land titles and improve their housing. Unfortunately, little of the money received seems to have been invested in tools or agricultural productivity.

Most of the outcomes hypothesized for departure areas, have been negative ones. The departure of younger, often better trained workers represent the loss of human capital that was formed at local expense. In so far as education is in part publicly supported, the community's investment in the migrant is not repaid, since he exploits it elsewhere. One of the contradictions of providing education in less prosperous regions of developing countries as a way of promoting regional development, is that the programme may simply be training people to migrate elsewhere. The region itself may benefit little from the educational programme.

As the young and potentially better qualified members of the labour force are drawn away, the work force left behind tends to be relatively older, less educated and less adaptable to new technologies and productive methods. An important hypothesis to be investigated is that, controlling age and education, those who initially migrate may be among the more innovative members of the community. Loss of leadership potential in rural communities is hard to assess in quantifiable terms, but the overall picture suggests that rural communities which lose their more skilled, innovative members may be less efficient, less adaptable, and ultimately less able to maintain the workers who remain. Morrison (1973, page 24) has referred to this possibility as a 'reverse multiplier' effect.

The above argument suggests that the heavy focus of research on migrants in urban areas may have been somewhat misplaced in so far as development related priorities and problems are concerned. The possibly negative impact of out-migration on the rural communities should not be ignored in developing nations which are dependent on agricultural production yet seem to have very serious problems in maintaining, much less increasing, agricultural productivity. An important research priority therefore, is the examination of the impact of out-migration on rural communities.

SUMMARY

In this paper we have sought to point out that research on migration in developing countries is in a transition stage. Much of the previous research has been descriptive of the characteristics of migrants, particularly poor migrants and squatters in large urban centres. In addition, there have been several studies based on census data, which give a broad picture of the socio-economic context of migration but little detail on the nature of the migrants themselves or on the characteristics of their native and receiving communities. Neither approach, therefore, has related migrant characteristics adequately either to the process of social change or to the development context in which it takes place. The absence of such information greatly limits the utility of existing research for policy and planning purposes. In seeking to outline the directions which future research should take, it was noted that some of the evidence available suggests that the age, sex, and occupational-skill characteristics of migrants may be interpreted within the context of a changing opportunity structure over time. Many specific questions and hypotheses are suggested by these data, but in general the research on this topic is at an early stage. It was also noted that the impact of migration on the communities involved, particularly on the areas of rural out-migration, has scarcely been examined yet evidence on this matter would seem to be essential for guiding policy action in the areas of urbanization, regional growth, and rural development. In conclusion, less attention should be given to descriptive research on the characteristics of migrants and the migratory process, and more attention should be given to the investigation of analytic hypotheses on the ways in which selective migration affects, and is affected, by the development process.

ACKNOWLEDGEMENTS

The author would like to acknowledge his gratitude to Marco Antonio Gramegna, Emanuel Jiménez and Abou Nabe who, as graduate students and as employees at IDRC in the summer of 1974, contributed to discussions on the ideas and bibliography presented in this paper.

NOTES

1. The following reviews of migration in selected regions are useful in so far as they consider population movement in the context of socio-economic change.
   AFRICA: Byerlee and Eicher (1970), Gugler (1969, 1974) and Hance (1970)
   ASIA: Eames (1969), Laquian (1972) and McGee (1971)
   LATIN AMERICA: Butterworth (1971), De Oliveira and Stern (1972), Gilbert (1974) and Cardona and Simmons (1974)
   WORLDWIDE: There are few cross-regional reviews, even on selected migration topics, although Breese (1966, Chapter 3) is useful on migrant characteristics, and Brigg (1971) is good in placing migration in an economic perspective. Kosiński and Prothero (1975) list 39 bibliographies of migration literature, several of which include research on developing countries.

2. In fact an increasing number of developing nations have such legislation even though it is typically poorly integrated with other development plans, and often not implemented at all. See Pryor (1974) for a discussion of some of these issues.
3. For a more detailed discussion of the relationship between public service costs and size of city see Chapter 7.

REFERENCES

Abrams, C. (ed.) (1966) *Squatter settlements: the problem and the opportunity* (Ideas and methods of exchange, No. 63, 302. Urban planning: Washington, D.C.: Office of International Affairs, Department of Housing and Urban Affairs), April.

Acosta, M., and Hardoy, J. E. (1972) 'Urbanization policies in revolutionary Cuba', in G. Geisse and J. E. Hardoy (Eds.), *Latin America Urban Research*, volume 2 (Beverly Hills: Sage Publications), 167–178.

Adams, D. W. (1969) 'Land ownership patterns in Colombia', *Inter-American Economic Affairs*, **17**, 77–86.

Alers, O. and Applebaum, R. P. (1968) 'La migración en el Perú: un inventario de proposiciones', *Estudios de Población y Desarrollo*, **1**(4).

Balán, J. (1969), 'Migrant-native socio-economic differences in Latin American cities: a structural analysis', *Latin American Research Review*, **4**, 3–29.

Balán, J., Browning, H., and Jelin, E. (Eds.) (1973) *Men in a developing society: geographic and social mobility in Monterrey, Mexico* (Austin: University of Texas Press).

Banks, J. A. (ed.) (1954) *Prosperity and parenthood* (London: Routledge and Kegan Paul, Ltd.).

Bantan, M. (1957) *West African city: a study of tribal life in Freetown* (London: Oxford University Press).

Bazán, C. (1975) *Family formation during a period of structural change: a study (proposal) in recently established Peruvian sugar cooperatives* (Lima: Centro de Estudios de Población y Desarrollo).

Bogue, D. J. (1969) 'Migration: internal and international', Chapter 19 in D. Bogue (ed.) *Principles of demography* (New York: John Wiley and Sons, Inc.).

Bogue, D. J., and Zachariah, K. C. (1962) 'Urbanization and migration in India', in Roy Turner (ed.), *India's urban future* (Berkeley: University of California Press).

Breese, G. (ed.) (1966) *Urbanization in newly developing countries* (Englewood Cliffs: Prentice Hall Inc.).

Brigg, P. (1971) *Migration to urban areas* (Economic Staff Working Paper No. 104; International Bank for Reconstruction and Development, International Development Association).

Browning, H. L. (1958) 'Recent trends in Latin American urbanization', *Annals of the American Academy of Political and Social Science*, **316**, 111–120.

Browning, H. L., and Feindt, W. (1969) 'Selectivity of migrants to a metropolis in a developing country: a Mexican case study', *Demography* **6**, 347–57.

Bruner, E. M. (1970) 'Medan: the role of kinship in an Indonesian city', in William Mangin (ed.), *Peasants in cities* (Houghton-Mifflin), 122–134.

Butterworth, D. (1971) 'Migración rural–urbana en América Latina: el estado de nuestro conocimiento', *América Indígena*, **XXXI**, 85–105.

Byerlee, D., and Eicher, C. K. (1970) *Rural employment, migration, and economic development: theoretical issues and empirical evidence from Africa* (Paper No. 1; African Rural Employment Study; Michigan State University; Department of Agricultural Economics).

Caldwell, John C. (ed.) (1969) *African rural-urban migration: the movement to Ghana's towns* (New York: Columbia University Press).

Cardona, R., and Simmons, A. B. (1974) *Hacía un modelo general de la población* (Bogotá: Antares and CCRP).

Conning, A. (1972) 'Rural-urban destination of migrants and community differentiation in a rural region of Chile', *International Migration Review*, **6**, 148–157.

Cornelius, W. (1969) 'Urbanization as an agent in Latin American political instability: the case of Mexico', *American Political Science Review* **LXIII**, 833–857.

Cornelius, W. (1970) 'The political sociology of cityward migration in Latin America', in Francine Rabinovitz and Felicity Trueblood (Eds.), *Latin America Urban Research*, Vol. 1 (Beverly Hills: Sage Publications), 95–147.

Cotler, J. (1970–71) 'Political crisis and military populism in Peru', *Studies in Comparative International Development*, **6**, 95–113.

Cummings, H. (1975) *Implications of migration for regional development planning: Thailand, Indonesia, the Philippines* (Ottawa: International Development Research Centre).

Davis, K. (1962) 'Urbanization in India: past and future', in Roy Turner (ed.), *India's urban future* (Berkeley: University of California Press).

Davis, K. (1963) 'The theory of change and response in modern demographic history', *Population Index*, **XXIX**, 345–366.

de Graft-Johnson, K. T. (1974) 'Population growth and rural–urban migration, with special reference to Ghana', *International Labour Review*, **109**, 471–486.

De Oliveira, O. and Stern, C. (1972) 'Aspectos sociológicos de la migración interna', *Economía Política*, **IX**, 85–100.

Deshmukh, M. B. (1956) 'A study of floating migration', in R. B. Textor, P. N. Prabho, A. F. A. Husain and M. B. Deshmukh (Eds.), *The social implications of industrialization and urbanization, five studies of urban populations of recent rural origin in cities of Southern Asia* (Calcutta: UNESCO).

De Voretz, D. J. (1972) 'Migration in a labour surplus economy', *Philippine Economic Journal*, **11**, 58–80.

Dorner, P. (1964) 'Land tenure, income distribution, and productivity interactions', *Land Economics*, **XL**, 247–254.

Eames, E. (1967) 'Urbanization and rural-urban migration in India', *Population Review*, **9**, 38–47.

Elizaga, J. (1970) *Migraciones a las areas metropolitanas de América Latina* (Santiago: Centro Latinoamericano de Demografía).

Flinn, W. L. (1968) 'The process of migration to a shanty-town in Bogotá, Colombia', *Interamerican Economic Affairs*, **22**, 78–88.

Flinn, W. L., and Converse, J. W. (1970) 'Eight assumptions concerning rural–urban migration in Colombia: a three-shanty town test', *Land Economics*, **46**, 456–466.

Friedlander, D. (1969) 'Demographic responses and population change', *Demography*, **6**, 359–382.

Gaur, R. S. and Nepal, G. S. (1962) 'Causes and consequences of rural emigration in East Uttar Pradesh', *Journal of Social Research* (India), **1**, 143–154.

Geisse, G., and Hardoy, J. E. (eds.) (1972) *Latin American Urban Research*, volume 2 (Beverly Hills: Sage Publications).

Gilbert, A. G. (1974) *Latin American development: a geographical perspective* (Harmondsworth: Penguin).

Godfrey, E. M. (1973) 'Economic variables and rural–urban migration: some thoughts on Todaro's hypothesis', *The Journal of Development Studies*, **10**, 66–78.

Gould, W. T. S., and Prothero, R. M. (1975) 'Space and time in African population mobility', in Leszek Kosiński and R. Mansell Prothero (Eds.), *People on the move* (London: Methuen & Co.), 39–49.

Gugler, J. (1969) 'On the theory of rural–urban migration: the case of sub-Sahara Africa', in J. A. Jackson (ed.), *Migration* (Cambridge University Press), 134–55.

Gugler, J. (1974) *Migrating to urban centers of unemployment in tropical Africa* (Paper presented to the Eighth World Congress of Sociology, University of Toronto, Toronto, Canada).

Hance, W. A. (ed.) (1970) *Population, migration and urbanization in Africa* (New York: Columbia University Press).
Harris, J. R., and Todaro, M. P. (1970) 'Migration, unemployment and development: a two sector analysis', *The American Economic Review*, **50**, 126–142.
Hauser, P. M. (ed.) (1961) *Urbanization in Latin America* (Paris: UNESCO).
Herrick, B. (1965) *Urban migration and economic development in Chile* (Cambridge, Massachusetts: MIT Press).
IDRC/Intermet (International Association for Metropolitan Research and Development) (1973) *Towndrift: social and policy implications of rural–urban migration in eight developing countries* (Ottawa: International Development Research Centre).
Jansen, C. J (1970) 'Migration: a sociological problem', in C. J. Jansen (ed.), *Readings in the sociology of migration* (London: Pergamon Press), 3–35.
Johnson, G. E., and Whitelaw, W. E. (1974) 'Urban–rural income transfers in Kenya: an estimated remittances function', *Economic Development and Cultural Change*, **22**, 473–479.
Kim, Y. (1972) 'Net internal migration in the Philippines', *Journal of Philippine Statistics*, **23**, 9–24.
Kosiński, L. A., and Prothero, R. M. (Eds.) (1975) *People on the move* (London: Methuen & Co.).
Laquian, A. A. (ed.) (1972) *Slums and squatter settlements in six Philippine cities* (New York: The Asia Society).
Lee, F. S. (1966) 'A theory of migration', *Demography*, **3**, 47–57.
Leeds, A., and Leeds, E. (1970) 'Brazil and the myth of urban rurality: urban experience, work, and values in "squatments" of Rio de Janiero and Lima', in A. J. Field (ed.), *City and country in the Third World* (Schenkman), 229–285.
Lowder, S. (1970) 'Lima's population growth and the consequences for Peru', in B. A. Roberts and S. Lowder (eds.), *Urban population growth and migration in Latin America: two case studies* (Liverpool: University of Liverpool, Centre for Latin American Studies), 21–37.
Mabogunje, A. L. (1970) 'Systems approach to a theory of rural–urban migration', *Geographical Analysis*, **2**, 1–18.
McGee, T. G. (1971) *The urbanisation process in the Third World* (London: G. Bell and Sons).
McGreevey, W. (1968) 'Causas de la migración interna en Colombia', in CEDE (1968) *Empleo y Desempleo en Bogotá* (Bogotá: Universidad de Los Andes), 211–221.
Mangin, W. (ed.) (1970) *Peasants in cities* (Boston: Houghton-Mifflin Co.).
Morrison, P. A. (ed.) (1973) *Migration from distressed areas: its meaning for regional policy* (Santa Monica: Rand Corporation).
Prachuabmoh, V., and Knodel, J. (1974) 'The longitudinal study of social and demographic change in Thailand: review of findings', *Asian Survey*, **14**, 350–364.
Pryor, R. J. (1974) *Population redistribution and development planning in South East Asia* (Paper presented to the Regional Conference of the International Geographical Union, Massey University, Palmerston North, New Zealand).
Pryor, R. J. (1975) 'Migration and the process of modernization', in Leszak A. Kosiński and R. Mansell Prothero (Eds.), *People on the move* (London: Methuen & Co.), 23–38.
Pye, L. W. (1963) 'The political implications of urbanization and the development process', in *United Nations Conference on the Application of Science and Technology for the Benefit of the Less Developed Areas, Geneva, 1962, United States Papers* (Washington, D.C.: U.S. Government Printing Office), 84–89.
Rehovot, Settlement Study Centre (1971) *Rural Urban Migration in Israel* (Jerusalem).
Roberts, B. R. (Ed.) (1973) *Organizing strangers: poor families in Guatemala City* (Austin: University of Texas Press).
Sahota, G. S. (1968) 'An economic analysis of internal migration in Brazil', *Journal of Political Economy*, **76**, 218–243.

Schultz, T. P. (1969) *Internal migration: a quantitative study of rural-urban migration in Colombia* (Santa Monica, CA.: Rand Corporation).

Schultz, T. P. (1971), 'Rural–urban migration in Colombia', *The Review of Economics and Statistics*, **52**, 218–245.

Simmons, A. (1970) *The emergence of planning orientations in a modernizing community: migration, adaptation and family planning in Highland Colombia* (Ithaca, N. Y.: American Studies Program, Cornell University, Monograph No. 15).

Simmons, A. and Cardona, R. (1972) 'Rural–urban migration: who comes, who stays, who returns? The case of Bogotá, Colombia, 1929–1968', *International Migration Review*, **6**, 166–181.

Smith, M. L. (1973) 'Domestic service as a channel of upward mobility for the lower-class woman: the Lima case', in Ann Pescatello (ed.), *Female and male in Latin America: Essays* (Pittsburgh: University of Pittsburgh Press), 191–208.

Stanford Research Institute (1969) 'Costs of urban infrastructure for industry as related to city size: India case study', *Ekistics*, **28**, 168.

Textor, R. B. (ed.) (1956) *From peasant to pedicab driver*, Cultural Report Series, No. 9 (New Haven: Yale University Press).

Todaro, M. P. (1969) 'A model of labor migration and urban unemployment in less developed countries', *American Economic Review*, **59**, 138–148.

Todaro, M. P. (1971) 'Income expectations, rural-urban migration and employment in Africa', *International Labour Review*, **104**, 387–413.

Turner, J. F. C. (1969) 'Uncontrolled urban settlement: problems and policies', in G. Breese (ed.) (1972), *Readings on urbanism and urbanization* (Englewood Cliffs, N. J.: Prentice-Hall, Inc.), 507–534.

Visaria, P. (1972) 'The adoption of innovations in agriculture and population trends in India', Paper presented at a seminar on *Effects of agricultural innovation in Asia on population trends*, Manila, Philippines: Ramon Magsayay Award Foundation,

Weiner, Myron (ed.) (1975) *Internal migration policies: purposes, interests, instruments, effects* (Cambridge, Massachusetts: Migration and Development Study Group; Center for International Studies, Massachusetts Institute of Technology).

Withington, W. A. (1967) 'Migration and economic development: some recent spatial changes in the population of rural Sumatra, Indonesia', *Tijdschrift Voor Economische en Social Geografie*, **58**, 153–163.

World Bank (1972) *Urbanization sector working paper* (mimeo.).

Chapter 4

The Role of Small Service Centres in Regional and Rural Development: With Special Reference to Eastern Africa

D. C. Funnell

INTRODUCTION

Most East Africans live in dispersed homesteads and derive a living from various forms of agriculture. The dispersed settlement pattern is punctuated by distinctive nodes which are sometimes large enough to be called towns but which are more usually only a small collection of buildings. These places are described as service centres in this essay, being the sites of medical, commercial, administrative and associated activities. It is not really appropriate to use the terms hamlet or village or call these places urban with the possible exception of the larger settlements. The more neutral term, service centres is preferable.

The growth of the large African cities such as Lagos, Nairobi and Kinshasa has been relatively fully documented. They are the foci of the nation's political and economic organizations whilst within them new social relations are being established. Despite the rapid growth of these cities it can be argued that they impinge directly on only a small proportion of the population. In Africa the level of urbanization is low, ranging from 30 percent in Ghana to the very low levels of 8–12 percent in East Africa. Even these figures are subject to debate due to the difficulty of deciding upon the most appropriate definition of an 'urban' centre.

Far less is known about the social, economic and geographical relations of the many small centres with which the majority of the population have more direct contact. The emphasis is on the local scale of these centres and Figure 4.1 provides an example of the distribution of service centres in one district of Uganda. In some countries these are included in the 'urban' category, elsewhere they are ignored. The only conceptual recognition of these places appears to be due to Middleton (1966, page 31). These centres, he argues, are the main loci for the dissemination of external influences to the majority of the populace and the recruiting ground for the incipient elites.[1]

Figure 4.1 Teso District 1970

These small centres are not just of academic interest. In recent years the increasing attention directed to regional planning has led to various proposals for the decentralization of services and industries. This has awakened greater interest in the potential significance of these small centres. As will be noted in detail later, the current Kenyan Plan (1974–1978) and Tanzanian Plan (1969–1974) incorporate specific proposals for the spatial organization of these smaller settlements.

The significance of these settlements in the overall pattern of urbanization and the space economy can be seen from Figure 4.2 which shows the city rank-size distribution for Uganda in 1969.[2] Primacy is evident, as it is in the other two East African countries, and the most striking feature is the dearth of centres of medium size. In all three countries, the capital cities are of the order of 250,000–500,000 people whereas there are very few places in the range 50,000–100,000. Consequently, in terms of regional policy and the potential for decentralization, the existing structure consists of many settlements ranging from around 2,000 persons to 20,000 persons. In certain areas these centres are the largest nucleated settlements.

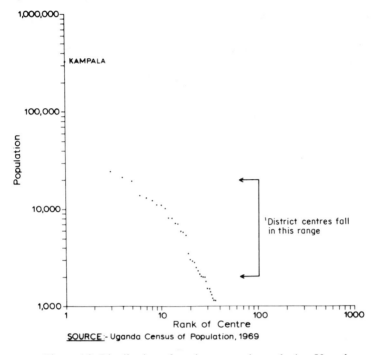

Figure 4.2 Distribution of service centres by rank-size, Uganda

Nevertheless, these small centres constitute the structure on which the plans have to be built. Several key questions arise which are significant in determining the suitability of various alternative strategies.

1. What regional or spatial pattern of centres is most appropriate for national policy requirements?
2. Which particular centres should be chosen for investment?
3. What kinds of investment are most appropriate for regional and rural development?

These are very difficult questions because not only do they require sophisticated analyses but also they raise important ideological issues. Indeed it is very difficult for practical proposals to be developed without clearly enunciated political goals. The close interrelations of the various factors means that certain priorities have to be stated very clearly. This is the prerogative of the politician and raises questions about the ideological position of planners and their relations with the decision-making process. This theme is not pursued here but is of obvious general relevance to any issue that involves resource allocation.

These service centres are involved in discussions of planning policy in several ways. They may be developed as a framework for the delivery of social services which ensures a reasonably equal distribution throughout the country, or sets of centres may be chosen to act as places through which new ideas are to be channelled in the hope that the innovations may stimulate change in the region around the centre. Finally, these settlements may be chosen as places for the location of economic activity which has been decentralized in an attempt to control the growth rate of the capital or other large cities.

The effectiveness of any particular policy depends upon the way in which these small centres fit into the overall socio-economic organization of the country. This has determined the framework for a review of the material at present available on small centres. There are already several commentaries upon existing government policy which make further extensive consideration redundant (Safier, 1969; Taylor, 1974). The application of particular analytical techniques is perhaps considered more effectively at the general level (Scott, 1971), and there have been few attempts to utilize sophisticated allocation–location models in the African context (Hirst, 1973).

THE EVOLUTION OF THE SMALL SERVICE CENTRES

The traditional socio-economic system has been destroyed by the penetration of capitalist modes of production. Consequently there has been a distinctive change in the geography of the countries. One element of structural change is the development of these small service centres. The penetration of capitalist activities came with the European colonial system and the critical factor in much of East Africa was the pressure to organize the extraction of a cash surplus from the rural economy. The economic and geographical structures that were developed linked the producer on a small holding with a consumer usually in Europe or America, and were controlled from London or Paris. Political control led to the establishment of a bureaucratic administration and the spread of an hierarchically organized series of administrative posts. In places such as Buganda, this pattern merely reinforced the existing political machinery. Where segmentary political systems had operated the new government posts heralded an entirely new form of political apparatus. The basic economic and political relationships have been modified further by missionary activity; various religious groups have established schools and medical services which often form the basis of a nucleated settlement.

In many parts of the continent these changes were carried out by alien groups. Strictly, this definition includes not only Europeans, Asians and Arabs but also Africans who have migrated from other areas. The Ibo trader in Hausaland, or the Buganda administrator in Lango, Uganda are examples of this. In East Africa, the Asian community has dominated the commercial sector, with Europeans being particularly important in the organization of international business. In 1966, the Census of Distribution in Uganda demonstrated that over 80 percent of the wholesale trade was in the hands of the Asian community. European business interests controlled only 8 percent of the total number of enterprises in that sector although they accounted for 21.5 percent of the total receipts.

The spatial structure of the evolving economies mirrors this domination by alien interests. Prior to colonial rule there were few interregional trade relationships and those that did exist were destroyed by the subsequent reorientation of trade towards exports.[3] There was little chance of the development of independent transactions between regions as the service centres channelled the flow of commodities towards the capital city and ports. The historical and geographic evolution of the pattern of nucleated settlements leads to a consideration of the views concerning the role of service centres in the process of development. The view held by Johnson (1970) in relation to centres in India, emphasizes the positive, beneficial effect to the local economy emerging from the increased development of the centres themselves. In contrast, Frank (1971) and others maintain that these small service centres are merely the outposts of a system of extraction which results in the net transfer of wealth from the rural areas to the large cities and ultimately to the metropolitan centres in other continents. Obviously these different interpretations have far-reaching and differing implications for the design of regional and rural development policies.

THE INTERNAL STRUCTURE OF SMALL SERVICE CENTRES

Research into the internal structure of small service centres has varied from the essentially sociological (Gutkind, 1968) to descriptions of the geographical organization of a whole class of centres (McMaster, 1968). Much of the sociological and anthropological work is concerned with the problems of adjustment to new forms of social relations which arise from the growth of new types of political systems and the nucleation of social and economic activity. Geographers have been more concerned with the internal pattern of economic activities and the spatial structure of housing and utilities.

The overall East African pattern is summarized by O'Connor (1968) who describes centres on the basis of their economic functions. Most of the centres considered here are administrative although they contain certain trading and occasionally small light industrial activities. The Colonial District towns are the main centres of the districts or regions and are very different from the surrounding dispersed settlements. McMaster (1965) argues that these towns are important not because of their size but because of the role they exert on the rural areas. This issue forms the basis of an extensive discussion in this essay.

The morphology of these towns is distinctive and reflects the imposed socio-economy of the area. The administrative *boma* contains those functions necessary for the handling of the district affairs. In Uganda during the period up to the late 1960's it included offices for the district political representatives. This *boma* area is usually quite distinct from the area formerly called the *bazaar* and which is now the main business zone. There are usually one or two banks, offices of various professions and a large number of shops. These latter are sometimes of a specialist nature such as a photographers or stationers but 'general stores' or *dukas* predominate. Until recently, most of the shops in the business areas of these district towns have been run by Asians. Africanization has been rapid in the general trades but less so in the specialist lines. In Uganda, of course, all commercial establishments in these small towns are now operated by Africans. The *dukas* play an important part in the operation of the exchange economy. They usually have a wide range of stock from crockery through to tinned milk, cement and cosmetics. Some traders combine the functions of retailing and wholesaling. In the larger centres it is usual to have many tailors and bicycle repairers operating on the sidewalks outside the shops. Most of these district towns have an urban market providing a range of fresh vegetables, dried fish, meat and small craft goods. They are usually strictly controlled and provide a source of income for the local authority that is independent of the central government. These institutions are important for the development of local agricultural exchange but few have been studied in any depth. In addition, the district towns have important social facilities. Bars and occasionally a cinema cater for all members of the community whilst various racial and religious groups have often developed their own exclusive facilities such as the European sports 'Club' and the Asian social centre.

Many of these towns have planned 'industrial' zones which in practice contain small repair shops or warehouses (*godowns*). Housing is also zoned and usually bears witness to the stratified social and racial pattern of the colonial society. Large spacious houses and gardens, built for the colonial administrators, now house the rising African élite and any remaining expatriates. Specific areas were allocated to the Asians with their own social or community lifestyles expressed in their design of houses. Finally, and usually on the periphery of the township, are the crowded low income African 'quarters'. Many of these district towns, therefore, possess town plans in the sense that zones are designated in which certain activities are allowed to take place. The internal structure of the centre thus reflects the image of the 'appropriate' town design from the viewpoint of the (normally alien) planner rather than any indigenous evolved pattern.

The same 'imposed' structure is also found in the other small centres although they are rarely zoned. At each locality there is an administrative office and usually medical facilities nearby. The trading area is often clearly separated and consists of one or two Asian *dukas* and African shops. These are usually built of mud and wattle, sometimes with a corrugated iron roof. The stock carried by these shops is very small, often just matches, cooking oil, salt and cloth. A local carpenter, bicycle repairer or shoemaker may have a plot in the centre.

It is in these 'rural trading centres' that much of the African enterprise has developed. Many shopkeepers are part-time, establishing a shop merely to provide a supplementary cash income but not expecting to derive all their livelihood from trading. Additionally there are important political advantages to trading which arise from the prestige of operating a store. Vincent (1971) in her pioneering study of the small trading centre of Bugondo in East Uganda has shown that the greater part of political power and prestige still comes from the control of agricultural resources but that the local 'Big Men' sometimes secure a stake in the local bar or shop. The whole question of the rural capitalist and the development of trade has not received the attention that it deserves in the East African context. There have been some interesting studies made of the businessmen in West Africa (Hill, 1970) and some work in Kenya (Marris and Somerset, 1971) but the development of trade goes hand in hand with the growth of the small centres. Thus in Uganda, for example, the attitude of the government towards trading was that 'business properly begins in the shop, not in the market' (Good, 1970, page 119). This implies that the location of trading operations would be increasingly focused in the small centres. The situation in Tanzania is perhaps the most interesting on this issue precisely because it has not yet been resolved. Small business trading enterprises are not officially discouraged though it could be argued that it is precisely this kind of operation that leads to the development of specific class structures, against which the ideological stance has been taken.

The distinctive spatial separation of the various functions is typical of both the large and small centres. In the case of the latter it sometimes appears as if they are separate nodes (Kade, 1969). Thus the schools and missions developed by the Europeans tend to be located two or three kilometres from the Asian *dukas*, with the African local government post and the dispensary perhaps a further kilometre away. Although today all these functions may be Africanized the legacy of this spatial organization has important consequences for rural planning.

Few of these centres offer any major source of non-agricultural employment apart from trade. Indeed the national surveys of employment usually fail to disaggregate the data to the level of these centres. The 1966 Census of Distribution in Uganda noted that approximately two thirds of those employed in trade (about 47,000) worked in rural areas. There are some jobs connected with government activities, especially road maintenance and other utilities but few skilled jobs exist. Those which do occur are often filled by outsiders. In the larger centres hawking takes place but there is little data on the magnitude in terms of population of the centre. It is therefore not possible to talk about an unemployment problem in these centres. Perhaps the arguments put forward by Weeks (1971) are more appropriate at this level than for the larger cities for which his ideas were developed. Most 'unemployed' people seem to be merely hanging around the shops or the government offices and have not yet emerged as an unemployed proletariat.

As these centres were the product of an alien administrative system it is not surprising to find their internal structure and economic activities controlled by government ordinance from their initial development. The question of the Town Plan has already been noted, but Langlands (1969) has described in detail the effect of various sets of legislation upon the physical and social structure. The boundaries of trading centres are defined in great detail in the Uganda case. More controversial has been the application of building standards to both residential and commercial premises within these townships. The pattern of housing has been described already while in the case of shops it is debatable whether the rules concerning their construction have prevent Africans from setting up in commerce in these centres. The Mabury Report (1955) provides interesting evidence from Uganda of the effect of these rules which led to the uniform design of the *duka* building. At 1954 prices a shop built to the required standards would cost approximately 14,000 shillings whereas if constructed from mud and wattle materials it would cost 4,000 shillings.[4] Obviously these standards imply that the potential trader needs considerable capital resources. Furthermore, the concentration of the Asian population in the East African townships is not entirely a question of a distinct racial preference for trading. Legislation in each country had the effect of preventing their participation in agricultural enterprises.

Planning modifications in the spatial structure of these centres involves recognizing the inertia generated by planning legislation which has determined much of the internal structure of the centres. The dispersed nature of the facilities within the centres adds significantly to the costs of providing piped water supplies, telephone communication or electricity (Glentworth, 1970). In Kenya, the development of nucleated settlements is encouraged although the stigma of the enforced villagization of the *Mau Mau* period remains. Tanzania has of course placed considerable emphasis on the development of *Ujamaa* villages with the concomitant institutional arrangements of the Village Development Committees (Connell, 1972). It is therefore a major part of Tanzanian rural and regional policy to group together households and it is hoped the majority of the population will reside in some form of nucleated settlement by 1976.

A policy based upon the grouping together of households is obviously important for the economic provision of utilities. However, given rural population growth rates, the result may be less desirable than expected. The process of moving the population to villages could take a considerable time unless carried out by force. If the utilities are concentrated in these village centres then the policy is increasing the real income of the community who already live in villages but not reaching the remaining rural populace. The 1971–1976 Uganda Development Plan provides a useful example of this dilemma. Many of the small centres considered in this survey require better utilities to maintain even minimum health standards and to encourage African traders to set up business in the centre. Some 93.5 million shillings have been allocated to water and sewerage supplies for many of these little townships but at most this will affect perhaps less than 6 percent of the population.[5] Less than two-thirds of this sum has been allocated to borehole provision although a higher proportion of the population could be served in this way.

The social structure of these centres is dominated by traders and administrators and constitutes an elite in relation to the rest of the population. In the Tanzanian situation the *Ujamaa* villages are conceived not only as a device for villagization but also as a particular social or communal institution. Nevertheless, it should be appreciated that spatial allocations of the kind mentioned are often also allocations to specific social groups in the population. Therefore it is important to be aware of the aggregate effect of transfers of this kind in determining the overall distribution of real income.

For larger centres, with populations exceeding 5,000, the problem is the lack of capacity in the physical infrastructure. Almost any moderate scale investment in industrial plant tends to incur high overhead costs. Water and electricity requirements for industrial purposes very quickly exceed the capacity of a supply system built for domestic users. Whilst this problem is widely recognized in the African context there have been very few attempts to provide workable planning tools to compare alternative development projects in these circumstances (Safier, 1970). Indeed it is only in recent years that serious attention has been devoted to infrastructural costs in any analytical framework (Lichfield, 1970). One attempt to remedy this problem is discussed in a brief paper by Fergus and Okulo-Epak (1970). They formulate the problem of new physical infrastructural needs in terms of 'threshold theory' developed by Malisz (1969). This provides a framework for determining significant points in the growth process which require proportionally large investments. Undoubtedly there is scope for further investigation with this technique in order to bring together the physical design process with the economic evaluation of alternative strategies.[6]

As much of the internal functioning of these small centres depends upon the administrative facilities located in them it is important to recognize the consequences of changing the pattern of administrative organization. Ocitti (1966) has examined the case of Kitgum, a small township which has been through the process of growth and decline almost entirely related to the allocation and then the withdrawal of administrative offices. Much of the income received in these centres is derived from the central government and once this is stopped business stagnates. Tissandier (1972) in his studies of a small township in Cameroun makes the point that much of the business is merely the recycling of urban consumption. The temporal and spatial variability of administrative centres has been studied by Hirst (1971). Clearly, therefore, whilst there may be a strong case for administrative reorganization the geographical consequences in relation to the existing townships need careful consideration.

Thus the internal structure of the small service centres in East Africa is essentially the result of the process of colonial legislation and the dominance of administrative functions. It is only with the attempts at villagization that we may witness a major reorganization of the settlement pattern. New development requires improved physical infrastructure and also the physical integration of many of the existing facilities. Inevitably this creates a situation in which people living in these centres become relatively better off than those outside, at least in terms of certain utilities.

SERVICE CENTRES AND CENTRAL PLACES

The service centre is the point at which town and countryside, and town and city interact. The theory of 'Central Places' is considered to be relevant to the planning of centres of this kind. This theory is perhaps the most widely recognized of all spatial theories and requires that the settlements are examined not in isolation but as part of a total system of such places. Under certain limiting assumptions the theory provides a rationale for the size, number and spatial distribution of places offering services to a dispersed population. The explanation of the patterns that emerge from this theory are of importance in trying to understand how such settlements operate and in estimating their potential as instruments of regional policy. There are many theoretical discussions of alternative settlement models which are considered as part of central place theory, but here only issues relating to planning problems are considered. Good summaries and bibliographies of the wider literature are made by Berry and Pred (1965) and Andrews (1970).

In its classical form the Christaller model of central places depends upon the emergence of a competitive equilibrium over space (Christaller, 1966). Centres provide functions which cater for discrete areas; business enterprises require a large enough area from which to derive sufficient revenue. Thus an interlocking lattice of market areas emerges, the theoretical hexagonal system. The number and size of settlements depends upon certain restrictions imposed by this geometry. In the Christaller model the settlements are organized in an hierarchy with special characteristics. All activities found at the lower levels of settlement also appear at the higher levels; centres at the same level have similar sets of activities and the emergence of specific higher order functions in the hierarchy is an indication of the level of the centre. The number of settlements at each level in the Christaller model is fixed according to certain 'organizing principles'. On the other hand, an alternative formulation by Lösch (1954) is more flexible but generally requires more centres to serve a given area. The 'administrative principle' of Christaller has received very little attention in the literature although it may be of considerable interest in the circumstances of the colonial economic and political structures.

The main elements of central place theory have become embedded in regional analysis and are used increasingly as a framework for the planning of systems of settlement in rural areas. Within East Africa this is best illustrated in Kenya where a series of infrastructural studies have led to the design of a national strategy which is incorporated in the 1970–1974 Plan and continued in the current Five Year Plan (Taylor, 1974). Similar ideas have been considered by planning authorities in other East African countries but have not yet been embodied in a total settlement strategy. Although there are problems using this theoretical structure the ideas contained within the theory appear superficially attractive for designing settlement patterns. It is thus important to determine just how far the existing pattern of centres conforms to the assumptions of this theory.

Within the wider context of Africa Grove and Huszar (1964) have used this theory as a framework to study the towns of Ghana. The urban pattern in Ghana is different from that of Eastern Africa as over 30 percent of the population live in towns with more than 5,000 persons. The study is a deliberate attempt to employ central place concepts in the planning of regional settlement patterns. The procedures adopted for determining the spatial and hierarchical structures were those common to many studies of central place systems at that time and they have been discussed by Davies (1966). For policy purposes, Grove and Huszar advocated a 'filling the gaps' strategy. The hierarchical structure was first indentified and places were noted which lacked any facilities that were possessed by other centres of similar rank in the hierarchy. Investment was then to be apportioned among those centres which showed a deficiency in facilities in order to bring them in line with others of that rank in the hierarchy. Although this may seem an eminently reasonable and practical approach to the problem it does make several unwarranted assumptions about the nature of the system. It is not clear from this work for instance whether the existing hierarchy is the most efficient for the provision of the relevant services.

A more sophisticated study of central places is that by Abiodun (1967) in southwest Nigeria. In Ijebu province she found a five-level system conforming very closely with the main principles of the theory. In neighbouring Abeokuta no such pattern existed primarily due to certain historical circumstances.[7] This contrast within a limited area shows that it is dangerous to make generalizations across broad geographical areas in circumstances where cultural patterns vary markedly. This is very characteristic of Africa where existing nation states consist of congeries of individual societies only recently pushed together under one authority. Therefore it is unwise to devise a strategy which involves the uniform application of a single model of settlement organization.

Within East Africa the work by Ponzio and Kamalamo (1966) follows the basic methodology of Grove and Husar and subsequently there have been a large number of studies concerned with central place systems. Mascarenas (1971) has examined the national pattern in Tanzania. In Kenya, a series of studies by Taylor (1970, 1972) and Obudho (1970) have demonstrated the existence of a central place structure. Similarly in Uganda, Kade (1969) has examined central places in Mengo, Splansky (1971) has studied Ankole and Funnell (1974) Teso. Altogether one might expect a wealth of relevant data for planning applications.

Certain conclusions can be drawn from the plethora of studies in East Africa. Undoubtedly the small service centres are organized on an hierarchical basis. Specific centres have a distinctive range of functions and in many cases the detailed structure resembles that of the Christaller model. Some of the spatial patterns indicate distributional regularities based upon the level of functions provided (Kade, 1969). However, several of the studies make unjustified assumptions about the nature of the processes underlying the pattern of settlements. It is taken for granted that the observed hierarchy is the result of a process of spatial competition and the emergence of distinct trade areas around the centres, even though many of the basic assumptions of the theory cannot be easily verified (Webber, 1971). For example, it is difficult to confirm the existence of a competitive pricing structure. In these studies of East African systems, the findings depend heavily upon the interpretation of the hierarchical structures, and often do not provide much evidence of spatial behaviour. Intuitive guesses are required, therefore, in arguing about the relevance of central place theory to understanding the existing system of service centres in East Africa.

With little industry and a dispersed rural population the conditions would appear appropriate for the establishment of a classical central place system. The introduction of cash crops and the development of trade should form the economic basis of a spatial structure of this kind. In fact the evolution of these particular centres has already been described pointing out the important role played by the colonial administration and alien traders. The pattern has been imposed upon the local inhabitants rather than evolving from the particular nature of the local economy. The administrative framework is of considerable importance but although the Christaller administrative principle seems relevant here the behaviour of the traders and the local population is treated as if it conforms to the more common 'market model'.

As the monetary economy extends its domination, so the system of centres will perhaps grow to play an increasingly significant role in the local economy. This evolutionary model is proposed by Bobek (1969) and appears to fit other parts of Africa. Indeed, Kade (1969) argues that within Uganda the system is more developed in the relatively affluent southern part of the country. If such a process of development is realistic then the policy of 'filling the gaps' to encourage the emergence of a fully intergrated hierarchy may be justified.

In Kenya, however, Taylor (1972) argues that the central place structure in some areas is composed of two systems existing side by side. One system is dominated by Africans and consists of markets and small shops while the other system is mainly run by Asians and is characterized by the larger formal trading centres. The two systems do interact at certain points and as Africanization proceeds the separation between them is diminishing. Nevertheless for planning purposes it is a mistake to assume that they operate as one fully integrated system, with all levels of centre being used by all sections of the population.

There have in fact been few detailed studies of the evolution of these centres in relation to the overall development of the economy.[8] Even if the hierarchical structures exist, this may be more the result of the organization of administration and the regulations concerning trading activities than the incipient emergence of a central place type settlement system.

For planning purposes some kind of hierarchical system is necessary for the provision of social services to a dispersed population. This being so it is surprising that none of these central place studies attempts to examine the question of the efficiency of this hierarchy in relation to specific services. The Kenya Town Planning Department made preliminary studies of the settlement structures in that country and these provided a considerable amount of information concerning the existing spatial structure. However in the Kenya Plan (1970–1974) the decision to adopt a four-level hierarchy in preference to any other is not discussed. It is questionable whether the resulting strategy is any more than a crude rationalization of the existing structure. The appropriate 'scale' of services has to be matched with a geographical area on the basis of the most flimsy evidence. In the United Kingdom the problems confronting the Redcliffe–Maud Commission provide some measure of the difficulty in circumstances that would appear far more favourable than East Africa. For practical purposes almost all the discussion in Kenya took place within the confines of the existing administrative structures and this is common practice elswhere. Nonetheless there is a strong case for the detailed examination of alternative patterns, at least in academic studies. The number of levels and the point in a system at which a particular service appears all need careful consideration.

For convenience it is usual to adopt the Christaller model. Each grade or level of settlement is assigned a distinctive range of functions and a discrete hierarchy is built up with many small centres and a few large. It is argued that this model represents the best, or perhaps more accurately the simplest compromise between the different scale requirements of various social services and administrative functions. As the existing system is often taken as a guide to the efficiency of the structure it is evident that problems can arise. In many cases it is likely that the number of facilities required to provide a service differ from function to function. For the most part little is known about the economics of the various services especially in terms of their effectiveness for the users. Consequently a more flexible model, perhaps akin to the Löschian settlement system would be more appropriate as a framework for the provision of services. This possibility has not been investigated in any studies known to the writer.

Given the difficulties of gathering information it is perhaps unfair to criticize planners for their apparent wholesale 'conversion' to the concept of central place theory as a means of structuring the spatial pattern of service centres in a rural economy. The available evidence suggests that the existing structure resembles a Christaller type model in form if not in substance. Unfortunately much of the research has stopped at this point leaving planners to devise a workable system which fits into the existing spatial structure. This should not conceal the fact that the common choice of a 'filling the gaps' policy using a Christaller type model is based more upon expediency than upon any proven efficiency or relevance to overall national goals.

THE RELATIONSHIP BETWEEN SERVICE CENTRES AND THE RURAL POPULATION

The relationship between the centre and the surrounding countryside lies at the crux of strategies which utilize the centres for regional and rural development. Some writers consider these settlements to be 'parasitic', sucking the wealth from the countryside, giving nothing in return (Hoselitz, 1955; Barber, 1967; Frank, 1971). The precise meaning of this term is a matter of contention. Planners simply assume that the provision of a facility in a centre constitutes a net addition to the real income of the centre itself and the region supposedly served by the centre. Furthermore some facilities are thought to provide a stimulus for change in rural areas. Several different relationships are involved in this problem. These are separated in the following discussion in order to ease presentation but obviously, in practice the patterns are very closely related.

The supply of social services

Social services mainly consist of health and education facilities but in some cases they also include extension services, administration, legal and community services. Several technical devices have been suggested for the optimization of a whole set of facilities in relation to a given population distribution. These include linear programming (Gould and Leinbach, 1966) and, more interestingly, combinatorial programming (Scott, 1971). Unfortunately they have yet to be applied to allocation problems within East Africa. This is primarily due to the lack both of appropriate data and the necessary technical expertise. The fundamental questions of geographers tend to focus on the problem of spatial accessibility and in order to investigate this, information is needed on the pattern of movements to facilities.

The various development plans incorporate long term goals for the provision of health centres on a per capita basis. In Kenya, the 1970–1974 Plan envisaged a network of health centres on the bases of one for every 20,000 persons. At present the average is 1:65,000 with regional values ranging from 1:50,000 to 1:100,000. Similarly, the current Tanzanian Plan proposes a ratio of 1:50,000 for health centres and 1:10,000 for dispensaries. There are problems in using these data for comparative purposes as the definition of a particular medical facility varies between countries. If regional policy is aimed at equalization between areas on the basis of data of this kind it must be recognized that such data do not include any information about the spatial accessibility of the facility.

In order to minimize the problem of accessibility most services are allocated to administrative areas rather than to specific locations. A certain number of dispensaries or primary schools may be assigned to each administrative unit on the basis of the population/facility ratios noted above. When the available facilities are limited obvious difficulties arise. If the unit is allocated on the basis of a long term plan, immediate local inequalities may arise until all the remaining elements of the plan have been implemented. If an attempt is made to optimize the location with respect to current requirements, duplication or overlapping of facilities may ultimately occur.

Beyond the immediate pragmatic issues lies the question of the efficiency of the administrative framework as a means of allocating facilities. Using this framework reduces the detailed analysis required for each decision, but inevitably assumes that the size of each area and the transport system within it provides a reasonable level of accessibility. If the number of people in administrative areas is approximately equal throughout the country, then it is usually taken for granted that this allocation procedure leads to a reasonably equitable solution. Where the population density within an administrative territory is uneven or where low densities lead to extensive administrative units, acessibility problems can become acute.

Measurements of spatial accessibility are designed to find out if certain parts of a territory remain unserved. This classic geographical problem is fraught with difficulties because most available data are derived from actual rather than potential movements. Estimates using gravity type models may be useful but encounter conceptual problems and difficulties arising from the fact that the data are often not sufficiently disaggregated. The actual distances obtained in any particular study are not always of much help in providing useful standards. In most cases they merely reflect the existing spatial separation of facilities, but unfortunately they are often the only data available.

Empirical information on the movement of people to services is most often gathered from the records at the destination. In a survey of small health facili-

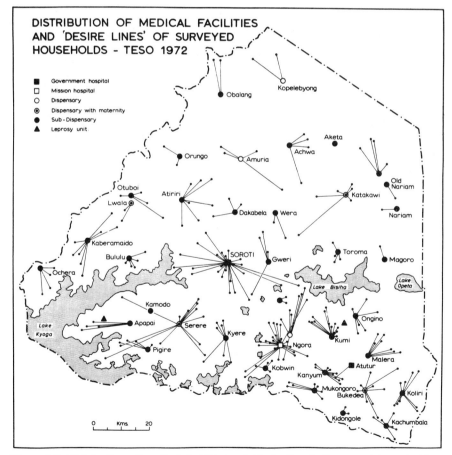

Figure 4.3 Distribution of medical facilities and 'desire lines' of surveyed households—Teso, 1972

ties in one district of Uganda (Funnell, 1974), a sample of male outpatients at each facility produced a list of 'addresses'. Characteristically these referred to the 'parish' which could be several kilometres in diameter. These data were supplemented by an attempt to determine the movement patterns from the point of origin, which was the household. In this case a sample of 250 was chosen using a stratified, random design. The information established that most households used the nearest available facility and in the area studied most journeys were less than 8 kilometres. The patterns are shown in Figure 4.3.

These data can be compared with the actual distribution of population. Straight line distances are measured between the mid-point of each parish to the nearest medical facility. Using the census data for the parish units it is possible to work out the proportion of the population that lives within certain distance bands of the various centres. Figure 4.4 shows that approximately 75 percent of the population lies with 9 kilometres of a medical facility. Although these measurements employ 'desire-line' data, this is not a major weakness in this particular area as there is a dense network of local tracks. The results correspond broadly with the actual movement patterns derived from the samples. Note, however, the relatively steep curve over the first few kilometres and the flatness beyond the 75 percent level. Approximately one-quarter of the population have to travel distances of up to 15 kilometres for even the most rudimentary treatment. Clearly this presents no mean task when the majority have to walk to these facilities. A more detailed study reported in King (1966) of a particular hospital also placed considerable emphasis on the question of accessibility. In this case, both in patients and outpatients were considered and again a very sharp distance-decay function was found outside the 8 to 12 kilometre band. Yet this hospital was supposed to serve a district which has a diameter of over 50 kilometres. Similarly, work by McGlashan (1972) in Malawi was aimed at establishing the patterns of usage in relation to the estimated capacity of the facilities concerned. He too found

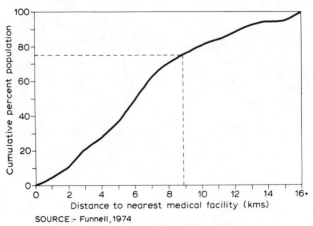

SOURCE :- Funnell, 1974

Figure 4.4 Distances travelled by cumulative proportion of population to medical facilities

wide discrepancies in the pattern of provision from the point of view of accessibility. Thus despite the common use of the referral system designed to minimize the costs of provision but maximize the availability of the appropriate treatment there is clear evidence of inequalities between certain areas and for certain people. In particular the very sharp distance-decay functions point to the fact that the centres in which these facilities are located effectively provide for only a limited proportion of the surrounding population.

Some similar problems have been observed with respect to education (Gould, 1972). In this case policy decisions create specific complications. In the case of secondary boarding schools, pupils may have to move relatively long distances at their own families' expense. Mukasa-Kintu's (1972) study in south Mengo, Uganda, shows how the religious bias of the school rather than simple proximity factors affects the choice for individual pupils. Hence inequalities may occur if certain religious groups are not represented locally. In Madagascar, Portais (1972) demonstrates how the proximity to secondary schools in the local township affects the proportion of pupils actually using them. It was found that 25 percent of the town children eligible for school actually attended school. At a distance of 10 kilometres from the township the proportion had fallen to only 9 percent.

The problem of accessibility needs to be considered more carefully within the constraints of capital and manpower. It is important to evaluate some of the more subtle influences on the use of facilities such as income and education. Unfortunately, census information rarely provides data on these matters. It may be argued that a sick person will travel long distances to seek a cure so that these finer details are relative luxuries in the circumstances of the East African economies. However, apart from raising the problem of the relationship between usage and personal wealth, such an attitude is inappropriate where the actual service being offered professes to be of a preventive nature. In this case the population has to be encouraged to use a particular facility and in this, spatial accessibility clearly plays an important role. Nevertheless the use of measures of accessibility based upon the attendance at facilities puts far too great an emphasis on the existing usage pattern as an indicator of service provision. The analysis is generally 'supply biased' with only very crude measures of demand being employed. Comparatively little work exists concerning the spatial pattern of disease or the variations in nutrition and vital demographic statistics. One example of the geographical approach to this problem is provided by Hall and Langlands (1968) which examines the general distribution of certain diseases. Another study by Thomas (1972) demonstrates how it is possible to determine certain important demographic variables, in this case child mortality, at a local level and the potential exists for this kind of work to be more closely integrated with the supply problems of health facilities.

The supply of consumer goods and services

The process of economic change in East Africa has involved the penetration of a capitalist exchange economy into a subsistence economy. The peasant farmer has been encouraged to produce crops for cash sale in order to pay government taxes and by the opportunity to obtain certain manufactured goods. The rural population is therefore a source of revenue from taxes, a producer of a marketable commodity (especially for export) and a market for manufactured goods. The nature of this exchange relationship is determined largely by the relative prices of the products involved and also by the institutional structure of supply and collection of the goods. An attempt to evaluate the results of rural–urban differentials in the real returns to rural productivity in Zambia perhaps raised more problems than it resolved but undoubtedly indicated the fact that the estimated terms of trade between the two sectors moved against the rural population (Maimbo and Fry, 1971).

In the past most transactions were handled by aliens, but increasingly state institutions have taken over major channels of distribution. The supply of goods and services by private traders depends upon the existence of opportunities for a suitable level of profit. Where state institutions are involved there may be other more important criteria for determining which centres are allocated specific kinds of business. In some cases state enterprise may be complementary, so that its locational pattern fills in areas unserved by commercial enterprise. More often, however, the object of state intervention is specifically competitive with the ultimate objective of controlling the particular business sector. In this case facilities are located alongside those already provided by private business.

In an area dominated by agricultural exchange the pattern of commercial activities plays a vital role in structuring the space economy. The goods and services provided can act as an important incentive for stimulating change in socio-economic relations. One important relationship is the frequency with which certain types of centres are used. The data presented in Table 4.1 indicates the order of magnitude of the differences between centres in one area of East

Table 4.1 Frequency of visits to centres in Teso, Uganda (% Households)

Frequency Centre	Up to 1 week	1 week to 2 weeks	2 weeks to 1 month	Over 1 month	Never
Low	52.4	32.0	10.6	5.0	—
Middle	15.8	20.8	29.7	21.3	6.4
High	6.3	11.7	22.3	46.7	13.0

Source: Funnell, 1974

Notes: 1. Typical functions at:
  low order: General store, Butcher, Subdispensary, Bar
  middle order: General wholesaler, Port agency, Petrol station, Dispensary
  high order: Dry cleaner, Bank, Hospital, Electrical goods
2. Data are derived from a sample of 250 households conducted in 1972

Africa. The types of goods available at each level are also shown. The mean travel distances from households to these centres are 3.2, 9.4 and 22.3 kilometres respectively, and so it is possible to relate the frequency pattern with that of distance travelled. The data indicate a high degree of interaction with low and medium centres. Approximately 68 percent of the households had made a visit to these two categories of centre in the week preceding the survey. It is also worthwhile noting that in this specific area 13 percent of the population had never visited the largest centre in the district although it was less than 80 kilometres from most settlements. Although it is difficult to structure questions on the frequency of visits or purchase of goods, these data do indicate that there is a major discontinuity between the larger centres and all others in terms of interactions. It is probable that facilities which are confined to the major township but which are intended to encourage greater participation in the exchange system will have little success. Where para-statal agencies control the distribution of certain goods it is important that sufficient outlets are maintained to serve the rural population. Experience in East Africa has shown how the limited number of skilled African managers has sometimes limited the number of stores operated by state trading companies. Local distribution has had to be left in the hands of private traders.

There are several reasons why state trading institutions have developed. Control over distribution and trade is a significant aspect of the management of the economy within East Africa. However in the recent past the most important motive for state control has been the pressure for Africanization.

The patterns of commercial functions are closely bound up with sociological and political factors. Kabwegyere (1968) discusses the motives of Africans setting up stores in a small trading centre and Marriss and Somerset (1971) use Kenyan data to illustrate some of the business problems facing the new African traders. The sociological conditions which set the trader apart from the population are very easy to identify since racial distinctions exist. The antipathy towards the Asians in East Africa is a compound of racial and economic antagonism. However, even where Africanization has been extensive the activity of trading sets apart the trader from his fellows. The growth of this petty capitalist group, even though it is not yet distinctive as a class, presents considerable problems to strategies of social and regional equality. In Tanzania, attitudes to traders at this level are ambivalent. They undoubtedly perform a useful economic function, both with respect to the supply of goods and the provision of credit. Yet their existence and the consequent opportunities for individual economic exploitation conflict with a strict interpretation of socialist development principles. Kenya, however, sees the rise of the African shopkeeper and entrepreneur as a signal step in the process of Africanization and economic development. Thus whilst physical planners may talk in terms of the encouragement of commercial growth at these centres, the particular institutions that perform these functions may have an important effect on relationships with the countryside. If the growth of businesses or the arbitrary allocations of government create enmity between peasants and traders in these centres then all attempts at using these places as points of contact and stimulants to change in the rural economy may well fail.

Service centres and agricultural production

A more specific relationship concerns the impact of these centres on agricultural production. A service centre can perform three functions in this respect. It may:-
  1. act as a local market or collecting point for the produce of local farms destined for consumption within the region,
  2. act as a collection centre for exported produce, the beginning of a chain of movement from the farm to an overseas consumer, or
  3. provide specific agricultural inputs or services to encourage the rural population to introduce technical changes in production.

Johnson (1970) has argued that the existence of a small service centre provides a powerful stimulus to agricultural change. He suggests that it is important to provide market facilities or collection points at these small centres which are accessible to the peasant farmers. The accessibility problem is similar to that already observed for other facilities, but far less is known about the 'field effect' of these centres on agricultural production itself. It might be expected that the impact would decline with increased distance from these centres as the diffusion of new ideas weakened. A whole armoury of techniques exist for examining the diffusion of new ideas but very few have been applied in the African context (MacKenzie, 1971). However, two studies do examine the impact of towns but not within a distinctive 'diffusionist' mould. Brandt, Schubert and Gerken (1972) have studied the degree to which the industrial town of Jinja, Uganda has led to changes in the pattern of local agriculture. The town of Jinja is much larger than those considered in this essay but the findings can be compared with those of Portais (1972) for the small township of Ambalavao. The dominant food demand in Jinja is for a plantain, *matoke* which is also the staple food of the population in the surrounding area, but has had little effect on the production of this crop. Supply is easily increased to cope with the urban demand because production techniques are very simple. The Malagasy case is rather different, however, since Portais finds a strong distance-decay effect in the area surrounding the town. Many new products such as pork, vegetables, fruit and groundnuts have been introduced and these require both inputs and technical advice which are only available from the township itself. Consequently, the products appeared on farms within about 15–20 kilometres from the township but beyond this few farms produced the new crops. The town makes a significant demand for foodstuffs and merchants visit the villages to buy for the township and for further shipment.

The contrast between the results of these studies is surprising. The larger centre would appear to have had less impact than the smaller town. Obviously there are differences between the two economies. One such difference concerns the institutional arrangements for handling the bulk of the marketed cash crop. In Malagasy much of the produce sold for cash was bought by agents at the local markets in the small centres. The township therefore acts as a crucial node in the marketing process of just those crops which involve the exchange of large amounts of cash. In parts of East Africa, especially Uganda, a similar system operated for only a relatively short period of time following the establishment of cash crop production. The authorities rapidly assumed control of the marketing of the cash crops (coffee and cotton). Special collection points were set up which were not always related to the network of trading centres. Consequently, the flow of produce and the pattern of transactions has not been so closely related to the service centres.

RURAL TRANSPORT AND SERVICE CENTRES

Modes of communication and transport directly condition all the interactions considered so far. Very little material exists on the nature of interpersonal communication and so the emphasis is placed upon transportation. At a regional scale, road transport is by far the most important method of moving people and goods. Roads consist merely of dirt tracks, or of gravelled road that are subject to maintenance at infrequent intervals. Certain interregional routes may be bitumenized . During a wet season it is not unusual for some roads to become impassable.

The modes of transport used by the population provide a useful starting point for discussion. Table 4.2 shows data from a sample survey in East Uganda which is very similar to the pattern all over this part of the continent. The 'modal split' is related to the level of centres in the district. The dominant mode of transport to small and medium level centres is by bicycle or on foot; for the high order places most people rely upon bus services.

The use of buses or taxis is a function of their relative costs. Although rates may be published for public vehicles it is by no means certain that they are consistently applied. Good (1970) reports that the average rate in Uganda for buses was 7 cents per kilometre. In addition, there may be charges for the carriage of goods. Officially a bunch of bananas should be charged 50 cents. At the time these figures were current, this commodity might sell for between

Table 4.2 Modes of travel to different orders of centre—Teso, Uganda
(% Sample Households)

Order of Centre	Mean Travel Distance km.	Foot	Bicycle	Taxi	Bus	Other
Low	3.17	38.5	59.0	—	1.7	0.8
Middle	9.40	16.0	63.0	3.2	17.0	0.80
High	22.30	6.0	17.0	10.0	65.0	2.0

Notes: Data collected as for Table 4.1 Source: Funnell, 1974

3 and 5 shillings. Thus transport costs could be between 10 and 15 percent of the final price. Few people have access to private motorized transport and so in rural areas buses are very important. However, maintenance costs are high and reliability may be low. Nearer main roads therefore taxis provide a useful and faster service at rates which are not very much higher.

One of the most controversial questions concerning rural transport is the likely benefit arising from improved feeder roads. There have been many studies examining the feasibility of trunk road developments but far less attention has been directed to feeder roads. It is these roads which are of importance for the interconnection of local centres. McKay, Roneche and Goshi (1971) have noted the high developmental effects resulting from reduced road transport cost and easy access to fertilizers, insecticides and agricultural advice (page iii). In some parts of Tanzania they found that the movement of crops from farm to collecting centre is performed by truck, elsewhere bicycle transport is sufficient. The upgrading of roads should benefit those areas where lorry transport is common as this would immediately lower the costs of wear and tear on the trucks. If bicycles are used, the immediate effects would be marginal, but in the long run if the improvements led to the introduction of motorized transport then substantial net benefits might be achieved. This is particularly true if the gains from the production of crops are limited by the inability of the transport system to move all the season's produce to the market.

The expansion of cash crops such as tea, tobacco, and cotton has usually been associated with plans for the improvement of local road networks. This was the case in Eastern Uganda in the early nineteenth century and is the same in Kenya for the expansion of tea production. Thus with improved accessibility to markets and relatively lower transport costs the real incomes of those living in rural areas should rise. But decreased transport charges alone may not benefit the producer if the lower haulage costs are not passed on. Hawkins (1962, page 157) makes it quite clear that the distribution of benefits depends at least as much upon the structure of economic relationships as on the actual decline in costs. If the transport system is a monopoly there may be no direct means of controlling the distribution of benefits through competitive pricing. The transport haulier may maintain his price to the producer and increase his profits. In turn the producer may be unable to sell his goods in a competitive market, and there may be no incentive for him to change methods of agricultural production.

For some centres the changes in the road pattern may not always be beneficial. A new road may deliberately by-pass a smaller trade centre. Greater access to an area from outside may provide opportunities for the merchants in larger centres to capture the trade leading to a general decline in the status of the smaller centre. Several interesting cases reported from the coastal region of Tanzania show how smaller centres depend on poor roads for protection (Hawkins, 1965). When road conditions improve with the onset of the dry season the local traders lose to bigger but more distant enterprises. At the same time, an improvement in real income is itself likely to make households more mobile and ultimately shift the focus of trade to more distant centres which offer a wider range of services.

The movement of wealth

The argument so far has assumed that the net effect of small centres is normally beneficial. Even if the activities operating in these small centres affect a limited geographical area changes introduced through the medium of these centres should raise the living standards of the rural population. This argument has been the subject of considerable debate following the criticisms levelled at capitalist economic development by Emmanuel (1969), Frank (1971) and others. Frank argues that it is the fundamental nature of capitalist economic relations that causes underdevelopment.[9] The expropriation of surplus by the metropolitan economies takes place through a series of 'satellites' ranging from the small market centres through to the national metropolis (Frank, page 131). Unfortunately, detailed analysis and quantitative expression of these internal relationships have yet to be performed in a rigorous manner.

Only fragmentary, rather peripheral information exists on the movement of wealth within the African context. Indeed the definition of wealth is itself a treacherous topic and so only some general points are presented here to indicate directions in which more work is required. Portais (1972) notes that investment by rural families in education represents a transfer of wealth to the towns. The first stage consists of the payment of fees and support costs of the children, monies which are recycled in the township economy or which are transferred to larger urban centres. Secondly, the children tend to leave the area seeking jobs in the country's large urban centres. This is a drain in invested wealth from the point of view of the rural population. Hence the educational system and the process of migration must also be seen in terms not only of labour migration but also the movement of wealth.

Land transfers play a similar role where there is a free market for land. In parts of Malagasy children sell off their inherited land to urban based persons. In turn these do not re-invest in agriculture in the rural areas but use it for speculative purposes. Hence wealth is being transferred to a landed class of urban dwellers.

There has long been discussion about the role of traders in the transfer of rural wealth. Some points have been made earlier about the particular characteristics of these traders but here the emphasis is upon their role as rural capitalists. It is asserted regularly that the level of profit obtained by these traders is too high, or that they exploit the rural population. The charge becomes particularly virulent when the traders are aliens. The evidence on this point is however slim and not really convincing. It is possible that very large wholesale traders, do make considerable profits, but unlikely that small traders in remote areas make excessive profits. Hawkins (1965) reports margins of 300–400 percent on certain commodities in remote stores. However, critics must take into account, and realistically price the effects of low turnover, high transport costs and storage problems. Where aliens are involved it is relevant to note that they often derive all their income from trading.

The prices paid for agricultural produce are a constant source of conflict. The small centre is the point in the chain at which the producer comes face to face with the 'representative' of a national or world trading concern. Antagonism is not the prerogative of the private trader; where the government sets the prices for certain crops, with little chance of bargaining, the feeling of exploitation may be just as great. Profits made by traders represent the transfer of wealth from one group of individuals to another. An independent local trader may reinvest this profit in other activities or use it to purchase local products for consumption. The small trader in a service centre is likely to operate on this basis. However, if the trader is merely an agent of a wholesaler located elsewhere who supplies capital both for premises and stock, spatial transfer of wealth can take place. Similarly, a private trader could reinvest any personal savings elsewhere but perhaps the best example is the operation of government enterprises which channel profits from individual operations into the centre.

The transfer of wealth is an integral part of the theory of comparative advantage and factor flows. Surpluses should be transferred to areas where the investment can be most useful, measured normally by the level of interest or rate of profit. How far this corresponds with a socially desirable pattern of investment is clearly a matter of much debate. In the East African situation there is no doubt that any transfers of this type would flow towards the large cities and probably out of the country.

The mechanisms for the removal of wealth from the countryside cannot be considered in isolation. Much of this essay has been concerned with the provision of social services financed from taxation and other funds made available from the central authorities. Remittances may also be important in family incomes. The problem becomes one of trying to develop a system of local accounts and this is well beyond the scope of this essay. What is evident, however, is that a study of the relationships between these service centres and the rural population constitutes an important part of the equation of the transfer of wealth. It is not satisfactory to assume that by constructing an integrated system of service centres and by providing the institutional framework for trade, this will invitably lead to a rise in rural incomes. Service centres may provide many of the crucial elements of socio-economic change; they may also harbour the mechanisms which lead to a decline in the position of the rural population in relation to other sectors of the population.

THE SMALL TOWN AS A GROWTH CENTRE

The previous sections have examined the way in which small service centres or townships provide social and business facilities for rural areas. It has been argued that while these centres are primarily intended to serve the immediate locality their potential value in the diffusion of impulses of change should not be taken for granted. In Kenya and Tanzania an explicit 'growth-centre' policy has been incorporated into their national plans. This policy is directed towards the larger of the centres considered in this essay. The Kenyan policy is concerned with the provision of infrastructural facilities in townships outside Mombasa and Nairobi. In particular, Nakuru, Kakmega, Kisumu, Eldoret, Thika, Nyeri and Embu have been designated major growth centres. The centres are to have priority for the development of services and other public works. Their economic base is to be founded upon government services and regional commerce. Kisumu, Eldoret, Nakuru and Thika which already have some industrial enterprises will be promoted as manufacturing centres. How ever there is no explicit statement that deals with the allocation of industries to these centres. For the most part the government is hoping that private industrial enterprises will repond to the enhanced infrastructural provision by locating new plants in these centres.

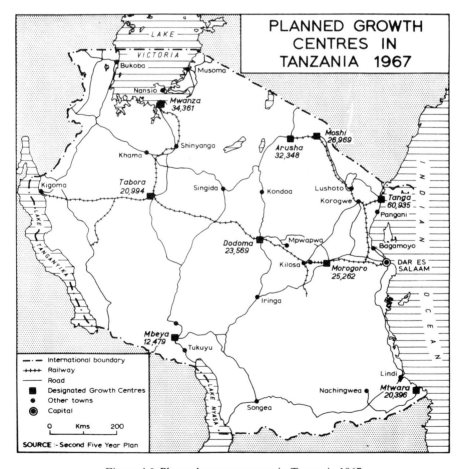

Figure 4.5 Planned growth centres in Tanzania 1967

In Tanzania the details of the regional policy of the Second Plan are discussed by Saylor and Livingstone (1971). Nine towns outside Dar es Salaam are to act as growth centres. These are Arusha, Dodoma, Moshi, Mbeya, Morogoro, Mwanza, Mtwara, Tanga and Tabora (Figure 4.5). Despite a considerable literature on growth centres it appears that little background analysis has been carried out prior to the announcement of these growth centres. Unlike Kenya however the Tanzanian government will take responsibility for allocating any new manufacturing investment to these centres.

Since the publication of the plan two useful studies have appeared which provide some background material for examining these growth-centre proposals. Hirst (1971) has examined the functional structure of the Tanzanian townships in terms of their occupational and employment data from the 1967 census. His findings confirm the general points made earlier about the uniformity of activities within these towns, for a relatively large proportion of the population is engaged in agricultural activities. Although some problems are caused by his particular techniques, he also finds there is little integration between the towns and their hinterlands. As the functional specialization is weakly developed there is very little interaction between these centres and most transactions are directed towards the capital city. Consequently, the development of viable production centres outside the capital requires a reorganization of the spatial structure and the encouragement of some measure of regional specialization.

Lundqvist (1973) describes the economic structure of one of the towns selected for investment, Morogoro. This town lies some 200 kilometres west of Dar es Salaam and has an estimated population of 32,000. Since Independence several new enterprises have developed in the town and it is argued that the nearby Tanzam railway should encourage its future expansion. This study makes a specific attempt to relate the activities within the town to the source of their inputs and the location of their outputs. Data on employment and the number of establishments is classified according to whether its inputs and outputs are 'local' or 'non-local'. The 'local' area incorporates the township and the surrounding Morogoro District. The data are set out in Table 4.3.

Although there may be some argument about the precise details of this classification the framework is very useful because it highlights just those features of the local economy that are important in growth-centre analysis. Much of the data was collected by survey though for certain categories Lundqvist was forced to rely upon the 1967 census. Hence occupations related to agriculture and marketing are probably underestimated.

Thus in Morogoro over 60 percent of the present employment produces an output which is consumed locally. Nearly half the inputs are derived from non-local sources. Unfortunately, it is not possible to express these data in value terms but it can be seen that non-local input makes up 46 percent and non-local output 38 percent of the employment. In this sense, it could be argued that the town is a net importer of jobs. If this is a correct reflection of the town's economic structure, then it is characteristic of a peripheral service

Table 4.3 Employment structure in Morogoro Town, Tanzania (% Total Enumerated Employees) (Reproduced by permission of the Scandinavian Institute of Afrikan Studies)

	Local Output	Non-local Output	
Local Input	45	9	54
Non-local	17	29	46
	62	38	100

Source: Lundqvist, 1973

Notes
1. Local Input/Local Output — markets and small scale processing, for example, charcoal burners
2. Local Input/Non-local Output — 'basic industries' processing raw materials from local area but selling product elsewhere
3. Non-local Input/Local Output — mainly retail trades
4. Non-local Input/Non-local Output — supplies purchased from outside region and only processed locally for further resale elsewhere. Also includes services with local offices in other parts of the country.

centre. This further emphasizes the magnitude of the transformation required to develop Morogoro into a regional growth centre.

It is important to relate this structure to the location-of-industry policy inherent in the Tanzanian growth-centre approach. In order to stimulate the local economy, it is necessary to develop activities in category 2 (and perhaps 4). Local resources need to be developed for products demanded in other areas. At present only 9 percent of employment is in this category. Yet, as Hirst has wryly commented the detailed criteria for determining the type of activity designated for these centres does not appear to have been worked out. All new industries which do not have to be located in Dar es Salaam can be allocated to these growth centres. Seidman (1972) has argued that a successful growth-centre policy requires careful selection of industries or activities involved. This implies that the industrial linkages are investigated with respect to both the existing industrial structure and any changes induced by the planned investment. The Morogoro study shows how only a small proportion of employment is based upon local inputs for non-local outputs. Also the largest manufacturing plant in the town, the tobacco factory, actually receives its main inputs from outside the district and most of the product is sent to Dar es Salaam. Therefore this manufacturing plant does not directly generate structural change in the local agricultural economy by backward linkages and the benefits only accrue through a weakened income and employer multiplier. It is not easy to investigate these multipliers in East Africa, but Lundqvist does point out that there are major leakages of local income due to the purchase of 'imported' manufactured goods.[10]

The survey of Morogoro reveals that there has been an upsurge in the number of enterprises established by African businessmen in the last few years. Many are small scale, often of the workshop type. This kind of activity could develop in response to the input requirements of the larger plants, for minor construction, transport or repairs leading to the development of local components. However, most of the existing larger plants depend upon components which are imported. No deliberate attempt is made to encourage the domestic or local production of these goods and consequently the local linkages are often not established. In fact, businessmen in Morogoro saw many opportunities for expansion and undoubtedly contribute to local employment and income expansion. The role of the small industries sector is a matter of current debate. It is usually labour intensive and tends to develop simple production techniques which are related to indigenous requirements. In western Nigeria, Bray (1973) has examined the opportunities for incorporating this level of activity into the spatial planning of the regional economy. This part of West Africa is however particularly rich in craft industries. In Eastern Africa, where interregional exchange has never been so extensive, there are some studies of the small industries but not in terms of their potential for the development of growth centres. Whilst these small industries create employment they generally pay lower wages than the larger manufacturing plants. Consequently, it is necessary to plan a collection of activities bearing in mind the balance between the effects of a larger volume of employment and few but relatively highly paid workers.

There are seventeen main towns outside Dar es Salaam and almost half of these have been designated growth centres. This is a high proportion when it is realized that most authorities argue that a successful growth centre must incorporate not just one but a linked series of activities which are able to establish and sustain a dynamic growth path. Saylor and Livingstone (1971) argue that if fewer centres had been chosen the chances of success would be proportionately higher. They suggest Moshi, Mwanza, Arusha and Tanga would be particularly suitable but these are just the centres which are comparatively well developed in relatively prosperous regions. Having chosen a larger number of centres in the interests of regional equity it may turn out that the growth-centre strategy fails and with it the possibility of achieving regional equity.

Alternatively, there is a case for encouraging regional specialization of the centres based upon the productive resources of the locality. Thus where certain agricultural inputs are available a linked sequence of processing plants could be established. This requires the clear ordering of regional priorities on the basis of a national plan for specialization. In Tanzania the plan does make some suggestions for this in relation to agriculture but does not really explore the possibilities in terms of the designated growth centres. However, it has been pointed out that the spatial structure of the Tanzanian economy exhibits an hierarchical structure in which there is little interaction between centres of the same level. In terms of a viable growth-centre policy it is undoubtedly necessary to induce more links between centres other than Dar es Salaam. This in turn means a careful analysis of the type of activity that should be encouraged in each particular centre.

The Tanzanian experience in regional planning is by no means unique and embodies the difficulties found in other less-developed countries. The two fundamental questions are firstly, which centres should be developed, and secondly, which activities should be encouraged? The planners are using the framework provided by the existing centres most of which exhibit the features discussed in this essay. The real danger is that, facing demands for apparent equity, the resultant allocation will merely consist of one 'modern' industrial plant for each regional centre. It is highly unlikely that such a pattern will induce substantive changes in the local economy or contribute to a dynamic growth in employment.

EXPERIMENTAL APPROACHES TO THE MODELLING OF SERVICE-CENTRE RELATIONSHIPS

Two problems have repeatedly arisen in this essay. Once concerns the design of an efficient spatial pattern of service facilities; the other the direct use of service centres as places from which new ideas and patterns of behaviour can be disseminated. Much of the current work on small centres is formulated in static terms, particularly that which relies heavily upon classical central place theory. There is a pressing need for concepts and techniques which match the processes of diffusion with the spatial structure of a particular area. Considerable information now exists on the nature of diffusion processes and in certain cases this has been specifically linked with the urban hierarchy (Pedersen, 1970; Berry, 1972). Unfortunately, the models employed in these diffusion studies cannot encompass a framework which is explicitly spatial. Furthermore, for practical purposes it is necessary to have a means by which alternative models can be tried out in the context of a specific geographical pattern of centres.

One of the most exciting possibilities lies in the application of network models to these problems. The potential scope of network analysis for geographical research is discussed in depth by Haggett and Chorley (1969) and several attempts have been made to use these techniques in the planning of urban development and road networks.

The procedures are initially simple. A set of service centres is represented as a matrix in which the elements define the nature of the links between the centres. In most of the experiments conducted so far the pattern of roads is used as an indicator of intercentre linkage. In some cases the mere existence of a link may be a satisfactory index but more sophisticated, weighted values could be employed. Tinkler (1972) has argued that certain mathematical operations on this matrix will simulate the likely flow of goods or ideas between the centres. This flow is partly dependent upon the accessibility of each centre relative to others. The resulting flows can be used to determine the probable distribution of the new innovation, or availability of goods after the process has reached some equilibrium stage.

The advantages of this kind of model lie in the fact that it is possible to alter the structural characteristics to experiment with different planning proposals. For example, different centres can be chosen through which a specific service or innovation is channelled. Similarly, the nature of the linkages could be changed, for example, by the upgrading of certain roads. The nature of the changes requires the construction of weights that are relevant to the specific circumstances of the area but generally do not affect the operation of the model itself.

There has been relatively little work on network models in Africa. Harvey (1972) has shown how certain concepts can be used to identify regional patterns of development while Smith (1974) demonstrates how vehicle flows between centres may be predicted. Funnell (1974) uses network analysis to model the effect of changing the point of entry of goods into a regional set of service centres. The results suggest that even simple assumptions can produce a pattern which closely resembles the distribution of functional status.

So far these techniques are experimental and there are several problems which need further investigation. In particular assumptions about the nature of the diffusion process need to be brought in line with current ideas of diffusion dynamics. Nonetheless, it is possible to manipulate the models so that the nature of the links between centres can be changed, or different centres proposed as the site of new facilities. In this sense this form of modelling may provide a useful tool for the operational comparison of different settlement patterns.

CONCLUSION

The development of small service centres in Eastern Africa is a product of the colonial administrative machine. They are important nodes in an otherwise dispersed population pattern. Although most commentators emphasize the hierarchical organization of the centres it must be remembered that this is a result of an imposed administrative pattern. It is common to employ Christaller's hierarchical concepts as a normative model of settlement structure, but there is surprisingly little evidence on the detailed spatial behaviour of the population.

The evidence demonstrates that the centres provide important social, commercial and administrative functions for only a limited proportion of the rural population. The distance-decay rates can be particularly sharp as most people are limited to travel on foot or by bicycle. Most planners are forced through expediency to utilize the administrative framework for the implementation of service provision. There have been few studies of the implications this may have on the spatial accessibility of facilities to various sectors of the population. Existing studies of service centres are singularly unhelpful as they generally end at the point of establishing a hierarchical organization. Much more work is required to link together spatial, social and economic factors to provide a suitable framework for service provision. The unadventurous nature of geographical work is startling in that few alternative spatial structures have been suggested in the light of differing needs of countries or regions. Whether a network approach assists in an examination of the alternatives remains to be seen, but certainly some evaluation methodology is required.

This remains particularly true when the notion of growth centres is introduced. In East Africa these have tended to be the large service centres at the upper end of the settlement hierarchy. Geographical studies do pinpoint the importance of establishing the pattern of linkages between the centres chosen and the countryside. Where no positive links exist or where no specific attention is directed as to how they develop, then clearly the growth centre is likely to be unsuccessful. Indeed, there is a grave danger that they merely act as a focus for government investment to service centres which themselves serve only a limited region. In this respect the centres would be more 'parasitic' than 'generative', reinforcing a pattern that some commentators consider typically colonial.

Finally, great stress has been placed on the need to consider the role of service centres in the light of the overall orientation of development policy in a country. There are dangers that the exercise is seen simply as one in which physical planners 'tidy' up or organize a set of centres according to notions of efficiency for service provision. Undoubtedly this is important, but the concentration on places rather than the people within them can have unfortunate long term consequences. Some geographical work, particularly by French scholars has stressed the equivocal nature of the role of service centres. A neatly integrated hierarchy may be logical for the downward diffusion of services and ideas, but it can also facilitate the rapid removal of wealth from the rural areas. Clearly this question involves much more than physical planning strategy. It is, however, intriguing to think that strategies devised to improve the real income of a local, rural population may be creating a structure which reinforces the existing inequalities between urban and rural areas.

NOTES

1. Middleton (1966) notes that the towns in East Africa had been classified into two types by Southall (1961). One was the traditional centre where the chief and his entourage settled; the other the modern colonial city, also the seat of an administration but this time alien. Middleton adds 'A third type of urban centre is the small trading and administrative centre' (page 33).
2. The interpretation of rank-size distributions has been the subject of considerable debate (Berry, 1961; Parr, 1970; Parr and Suzuki, 1973). The data is presented here for descriptive purposes only.

3. See Uzoigwe, G. N. (1971) and Wood, L. J. (1975).
4. In 1954 approximately 20 Ugandan shillings were the equivalent of £1–00 sterling.
5. In 1974 approximately 17 Ugandan shillings were the equivalent of £1–00 sterling.
6. Threshold analysis examines the pattern of investment in essential intrastructure required by the growth of a town. It is argued that costs rise in a discontinuous fashion so that there are certain jumps in the amounts required for expansion and improvement. These jumps are caused by the capacity constraints of water supply, roads, sewage works, etc. The identification of the 'thresholds' is said to assist in long term or strategic planning of a settlement.
7. Local warfare in the 19th century led to the population vacating their villages and moving into the major town of the area, Abeokuta.
8. Personal communication. Muwonge has recently completed a thesis on the genesis of central places in Mengo District, Uganda.
9. This is not the place to discuss the general merits of this argument. An interesting critique is provided by Nove (1974).
10. Studies in the United Kingdom and France by Moseley (1973) have indicated similar tendencies for industrial plants decentralized from the capital cities.

REFERENCES

Abiodun, J. O. (1967) 'Urban hierarchy in a developing country', *Economic Geography*, **43**, 347–367.

Andrews, H. F. (1970) *Working Notes and Bibliography on Central Place Studies*, 1965–1969 (Working Paper No. 8; Department of Geography, Toronto).

Barber, W. J. (1967) 'Urbanization and Economic Growth: the case of two white settler territories', *The City in Modern Africa*, Ed. H. Miner (London: Pall, Mall).

Berry, B. J. L. (1961) 'City-Size Distributions and Economic Development', *Economic Development and Cultural Change*, **4**, 573–587.

Berry, B. J. L. (1972) 'Hierarchical diffusion: the basis of development filtering and spread in a system of Growth Centres', *Man, Space and Environment*, Ed. P. W. English (London: Oxford University Press), 340–359.

Berry, B. J. L. and Pred, A. (1965) *Central Place Studies—a bibliography of theory and applications* (Regional Science Research Institute; Bibliographic Series No. 1).

Brandt, H., Schubert, B. and Gerken, E. (1972) *The Industrial Town as a Factor of Economic and Social Development* (Munich; I. F. O.).

Bray, J. (1973) 'Small Scale Industries and Urban Development in Western Nigeria' (*Institute British Geographers; Annual Conference 1973*, unpublished paper).

Bobek, H. (1964) 'Die hauptstufen der Gesellschafts und Wirtschaftsentfaltung in geographischer Sicht', *Die Erde*, **90**, 259–298.

Chorley, R. J. and Haggett, P. (1969) *Network Analysis in Geography* (London: Arnold).

Christaller, W. (1966) *Central Places in Southern Germany*, trans. W. Baskin (Englewood Cliffs, N. J.: Prentice Hall).

Connell, J. (1973) 'Ujamaa Villages: Institutional change in rural Tanzania', *Journal of Administration Overseas*, **12**, 273–283.

Davies, W. K. D. (1966) 'The ranking of service centres: a critical review', *Transactions, Institute of British Geographers*, **40**, 51–55.

Emmanuel, A. (1969) *L'échange Inégal* (Paris; F. Maspero).

Fergus, M and Okulo-Epak, F. (1970) 'A report on the application of urban threshold theory in Uganda', *The Role of Urban and Regional Planning in National Development in East Africa*, Ed. M. Safier (Kampala: Milton Obote Foundation).

Frank, A. G. (1971) *Capitalism and Underdevelopment in Latin America* (Harmondsworth: Penguin).

Funnell, D. C. (1974) *Service Centres in Teso District, Uganda* (Unpublished D. Phil. thesis, University of Sussex).

Glentworth, G. (1970) 'Rural electrification and planning—a Uganda case study', *The Role of Urban and Regional Planning in National Development in East Africa*, Ed. M. Safier (Kampala: Milton Obote Foundation).

Good, C. M. (1970) *Rural markets and trade in East Africa* (Department of Geography Research Paper No. 128: University of Chicago).

Gould, P. R., and Leinbach, T. R. (1966) 'An approach to the geographical assignment of hospital services', *Tijdschrift voor Economische en Sociale Geographie*, 57, 203–206.

Gould, W. T. S. (1972) 'Patterns of lower school enrolment in Uganda', *East African Geographical Review*, 10, 65–74.

Grove, D. and Huszar, L. (1964) *The Towns of Ghana: the role of service centres in regional planning* (Accra: Ghana University Press).

Gutkind, P. C. W. (1968) 'The small African town', *African Urban Notes*, 3, 5–11.

Hall, S. and Langlands, B. W. (1968) *Uganda Atlas of Disease Distribution* (Kampala: Department of Preventive Medicine; Makerere University).

Harvey, M. E. (1972) 'The identification of development regions in developing countries', *Economic Geography*, 48, 229–243.

Hawkins, E. K. (1962) *Roads and road transport in an underdeveloped country* (London: H. M. S. O.).

Hawkins, H. C. G. (1965) *Wholesale and retail trade in Tanganyika* (New York: Praeger).

Hill, P. (1970) *Studies in rural capitalism* (Cambridge: Cambridge University Press).

Hirst, M. A. (1971) 'The changing patterns of district administration centres in Uganda since 1900', *Geographical Analysis*, 3, 90–98.

Hirst, M. A. (1973a) 'A functional analysis of towns in Tanzania', *Tijdschrift voor Economische en Sociale Geographie*, 64, 39–51.

Hirst, M. A. (1973b) 'Administrative reorganization in Uganda: towards a more efficient solution', *Area* 5, 177–181.

Hoselitiz, B. F. (1955) 'Generative and parasitic cities', *Economic Development and Cultural Change*, 3, 278–294.

Johnson, E. A. J. (1970) *The organization of space in developing countries* (Cambridge: Harvard University Press).

Kabwegyere, T. B. (1968) *The growth of a trading centre in rural Ankole: Ishaka* (Department of Sociology: Working Paper No. 57; Makerere University).

Kade, G. (1969) *Die Stellung der Zentralen orte in der Kulturlanschaftlichen entwicklung Bugands* (Frankfurt: Frankfurter Wirtschafts und Sozial geographische; Heft 6).

King, M. (1966) *Medical care in developing countries* (Nairobi: Oxford U.P.).

Langlands B. W. (1970) 'Urban functions and urban forms in Uganda and their implications for planning policy', *Perspectives on urban planning for Uganda*, Ed. Langlands, B. W. and M. Safier, (Department of Geography: Occasional Paper No. 10; Makerere University).

Lichfield, N. (1970) 'Evaluation methodology of urban and regional plans: a review', *Regional Studies*, 4, 151–165.

Lösch, A. (1954) *The Economics of Location* (New Haven: Yale U. P.).

Lundqvist, J. (1973) *The Economic Structure of Morogoro Town* (Uppsala: Scandanavian Institute of African Studies; Report No. 17).

Mackenzie, M. K. (1971) *Present location and past diffusion of the cured tobacco industry in W. Nile* (Department of Geography; Occasional paper No. 21; Makerere University).

Mabury, M. A. (1955) *The Advancement of Africans in Trade*, (Entebbe: Uganda Protectorate).

Maimbo, F. T. and Fry, A. J. (1971) 'Investigation into the change in the terms of trade between rural and urban sectors of Zambia', *African Social Research*, 12, 95–110.

Malisz, B. (1969) 'Implications of threshold theory for urban and regional planning', *Journal of the Town Planning Institute*, **55**, 108–110.

Marris, P. and Somerset, A. (1971) *African Businessmen* (London: R. K. P.).

Mascarenas, A. C. (1971) 'Urban Centres', *Tanzania in Maps*, ed. L. Berry (London: University Press).

McGlashan N. D. (1972) 'Distribution of population and medical facilities in Malawi', *Medical Geography: techniques and field studies*, ed. N. D. McGlashan (London: Methuen), 89–96.

McKay, J. K., Roneche, K. and Goshi, T. (1971) *The feasibility and planning of road improvements in the cotton growing areas of Geita District* (Dar es Salaam: Bureau Resource and Land Use Planning; Paper No. 16).

McMaster, D. N. (1968) 'The colonial district town in Uganda', *Urbanization and its Problems*, Ed. R. P. Beckinsale and J. M. Houston (Oxford: Blackwells), 330–367.

Middleton, B. J. (1966) *The effects of economic development on traditional political systems in Africa south of the Sahara* (The Hague: Mouton).

Moseley, M. J. (1973) 'The impact of growth centres in regions, Part II. An analysis of spatial flows in East Anglia', *Regional Studies*, **7**, 77–94.

Mukasa-Kintu (1972) *Distribution of schools and patterns of movement to school in Mukono area, Uganda* (Department of Geography: B. A. Research Paper; Makerere University).

Nove, A. (1974) 'On reading A. G. Frank', *Journal Development Studies*, **10**, 445–455.

Obudho, R. A. (1970) 'The central places of Nyanza Province, Kenya: a tentative study of urban hierarchy in a developing country', *African Urban Notes*, **5**, 382–391.

Ocitti, J. P. (1966) 'Kitgum, an urban study', *East African Geographical Review*, **4**, 37–48.

O'Connor, A. M. (1968) 'The cities and towns of East Africa: their distribution and functions', *Ostafricanische Studien*, **8**, Ed. H. Berger, 41–52.

Parr, J. B. (1970) 'Models of city size in an urban system' *Papers and Proceedings Regional Science Assoc.*, **25**, 221–253.

Parr, J. B. and Suzuki, T. (1973) 'Settlement populations and the lognormal distribution', *Urban Studies*, **10**, 335–352.

Pedersen, P. O. (1970) 'Innovation diffusion in a national urban system: the case of Chile', *Geographical Analysis*, **2**, 203–254.

Ponzio, M. and Kamalamo, P. (1966) 'The application of central place theory in Mengo and Busoga districts, Uganda', *Papers, East African Social Science Conference, Geography Section*.

Portais, M. (1972) 'L'influence d'une petite ville sur son environment rural; le bassin Ambalavao', *La Croissance Urbaine en Afrique Noire et Madagascar*, Ed. P. Vennetier (Paris; C. N. R. S., 2 vols).

Safier, M. (1970) 'The development of urban infrastructure for industrial needs', *The Role of Urban and Regional Planning in National Development in East Africa*, Ed. M. Safier (Kampala: Milton Obote Foundation).

Saylor, R. G. and Livingstone, I. (1971) 'Regional Planning in Tanzania', *The African Review*, **1**, 53–70.

Scott, A. J. (1971) *Combinatorial programming: spatial analysis and planning* (London: Methuen).

Seidman, A. (1972) *Comparative Development Strategies in East Africa* (Nairobi: East Africa Publishing House).

Smith, J. A. (1974) 'Regional inequalities in internal communications: the case of Uganda', *Spatial Aspects of Development*, Ed. B. S. Hoyle (London: Wiley).

Splansky, J. B. (1971) *Emergent urban places in Africa: the case of Ankole, Uganda* (Unpublished Ph. D. University of California, Los Angeles).

Taylor, D. R. F. (1970) *Development of central places in Coast Province Kenya* (Ottawa, Carleton University).

Taylor, D. R. F. (1972) 'The role of the smaller urban place in development: a case study from Kenya', *African Urban Notes*, **6**, 7–23.

Taylor, D. R. F. (1974) 'Spatial aspects of Kenya's rural development strategy', *Spatial Aspects of Development*, Ed. B. S. Hoyle (London; Wiley) 167–188.

Thomas, I. D. (1972) 'Infant mortality in Tanzania', *East African Geographical Review*, **10**, 5–26.

Tinkler, K. I. (1972) 'The physical interpretation of eigenfunctions of dichotomous matrices', *Transactions Institute of British Geographers*, **55**, 17–46.

Tissandier, J. (1972) 'Aspects des relations ville campagnes dans le département de la Haute-Sanaga (Cameroun)', *La Croissance Urbaine en Afrique Noire et Madagascar*, Ed. P. Vennetier (Paris: C. N. R. S.; 2 vols.) 1029–1046.

Uzoigwe, G. N. (1971) 'Pre-colonial markets in Bunyoro-Kitara', *Comparative Studies in Society and History*, **14**, 422–453.

Vincent, J. (1971) *African Elite: Big Men in a Small Town* (New York; Columbia University Press).

Webber, M. J. (1971) 'Empirical verification of classical central place theory', *Geographical Analysis*, **3**, 15–28.

Weeks, J. F. (1971) 'Wage policy and the Colonial legacy: A comparative study', *Journal of Modern African Studies*, **9**, 361–387.

Wood, L. J. (1975) 'Glottonchronology and research in historical geography', *Area*, **6**, 251–253.

Chapter 5

Regional Income Disparities and Economic Development: A Critique

Alan G. Gilbert and David E. Goodman

It is ten years since the publication of Williamson's well-known article on the relationship between economic growth and regional income disparities (Williamson, 1965). During that time an extensive literature has appeared providing further empirical evidence and examining the processes involved in spatial differentiation. Despite this flood of writing, however, we are still far from agreement on a number of critical questions. Do regional incomes converge as per capita income rises? What are the most appropriate policies to combat regional income divergence? Should efforts be made to remove regional disparities at an early stage of development or be postponed until a country has achieved higher levels of economic development?

This lack of agreement is critical in so far as regional development strategies increasingly are being adopted by national governments (UNRISD 1970–1974; Gilbert, 1974). Both in the developed and the less-developed world, policies to modify the existing spatial distributions of economic activity have been introduced. Urbanization and regional imbalance are now regarded as matters sufficiently important to be included in national development plans. In Latin America, for example, Utria has recognized that 'like import substitution and industrialization in the fifties, and national planning and economic integration in the sixties, regional development appears destined to become one of the principal concerns of planners and strategists of Latin American development' (Utria, 1972, page 61). Similarly, Lasuén has noted that 'suddenly in the decade of the 70's without any transition ... the urban–regional theme in almost all countries has moved up the scale of national aspirations from the base to the top' (Lasuen 1974, page 89).

The reduction of regional income differentials is clearly one of the major goals of regional development planning. While other goals such as the wish to open up new land areas, integrate regions more firmly into the national economy and reduce the pressures building up in metropolitan areas also may be espoused, income equalization is often the key regional objective. Indeed, with development economists and international agencies concentrating increasingly on distributional and employment objectives, the regional dimension may well assume greater importance in economic policy in the future (ILO, 1970, 1972; Furtado, 1973; Sundrum 1972; UNECAFE, 1971).

In this context, it is important that we should seek to understand more fully the processes which lead to convergence and divergence in regional incomes. For, despite numerous empirical case studies, the massive literature on urbanization and primacy, and the extensive debate on growth centres and optimum city size, it is doubtful whether we have clear answers to a number of vital planning questions. Among these questions are two with which this paper is basically concerned. The first is whether the process of income convergence is as inevitable as Williamson's empirical findings appear to suggest. The second is to ask whether regional income equalization should constitute a major goal of development planning.

The paper is organized into four sections. The first examines Williamson's findings and the assumptions implicit in his approach. This section also supplements his work with regional income data which were not available in the early sixties. The second part of the paper examines the theories and arguments which have been put forward to explain regional income convergence and divergence. It also considers whether the processes which brought about convergence in the developed countries are likely to operate in the less-developed nations in the future. The third section examines the usefulness and relevance of the criterion of regional income equalization in development planning. In particular, it poses the question whether policies to encourage regional income convergence are compatible with, and identical to, strategies aimed at improving the distribution of personal income. How far indeed do regional policies help reduce personal disparities both within nations as a whole and within those nations' poorest regions? This last question is examined in detail in the final section of the paper drawing on the experience of Northeast Brazil. This case study is of special interest since Brazil, which is frequently cited as an example of severe regional 'dualism', has been experiencing rapid economic growth in recent years and has taken a number of important policy steps to reduce regional income disparities (Robock, 1963; Baer, 1964; Furtado, 1963; Hirschman, 1969; Goodman, 1972; Roett, 1972).

THE EMPIRICAL EVIDENCE

Williamson's article remains the most comprehensive study of regional disparities available. In that study, he examines the hypothesis that regional income disparities are related to national levels of economic development. On the basis of cross-section data for 24 countries, he finds that the nations with the largest regional differentials are drawn from a group with intermediate levels of per capita income, whereas highly developed nations and those which have experienced only limited economic growth exhibit relatively small regional disparities. (His indices of regional inequality are included in Table 5.1.)

Further support for this finding is provided by time-series data for ten countries. Williamson argues that 'what little information we have on nineteenth and twentieth century Italian, Brazilian, United States, Canadian, German, Swedish and French experience suggests that increasing regional inequality is generated during the early development stages, while mature growth has produced regional convergence or a reduction in differentials'

Table 5.1 Williamson's international cross-section

Country and Kuznets group classification	V_w	V_{uw}	Years covered
Australia	0.058	0.078	1949/50–1959/60
New Zealand	0.063	0.082	1955
Canada	0.192	0.259	1950–61
United Kingdom	0.141	0.156	1959/60
United States	0.182	0.189	1950/61
Sweden	0.200	0.168	1950, 1955, 1961
Group I average	0.139	0.155	
Finland	0.331	0.276	1950, 1954, 1958
France	0.283	0.215	1954, 1955/6, 1958
West Germany	0.205	0.205	1950–55, 1960
Netherlands	0.131	0.128	1950, 1955, 1958
Norway	0.309	0.253	1952, 1957–60
Group II average	0.252	0.215	
Ireland	0.268	0.271	1960
Chile	0.327	0.440	1958
Austria	0.225	0.201	1957
Puerto Rico	0.520	0.378	1960
Group III average	0.335	0.323	
Brazil	0.700	0.654	1950–59
Italy	0.360	0.367	1951, 1955, 1960
Spain	0.425	0.356	1955, 1957
Colombia	0.541	0.561	1953
Greece	0.302	0.295	1954
Group IV average	0.464	0.447	
Yugoslavia	0.340	0.444	1956, 1959, 1960
Japan	0.244	0.222	1951–9
Group V average	0.292	0.333	
Philippines	0.556	0.627	1957
Group VI average	0.556	0.627	
India	0.275	0.580	1950/51, 1955/6
Group VII average	0.275	0.580	
Total average	0.299	0.309	

(Williamson, 1965, page 44). From this evidence, he hypothesizes that as per capita incomes increase, relative regional disparities first widen, then remain steady and subsequently decline. While he questions the adequacy of his evidence to support the idea of initial divergence, he argues that both the time-series and the cross-national results '... suggest that there is a systematic relation between national development levels and regional inequality ... rising regional income disparities and increasing North–South dualism is typical of the early development stages, while regional convergence and a disappearance of severe North–South problems is typical of the more mature stages of national growth and development' (Williamson, 1965, page 44). It should be noted, however, that Williamson is careful to point out that convergence is limited to relative disparities and that absolute income differences may still increase over time.

Williamson's findings have been accepted by many writers and have been absorbed into the conventional wisdom of planning. They have also been linked implicitly, rather than formally, to the changing size distribution of cities over time (Friedmann, 1972–1973; Vapñarsky, 1969; Keeble, 1967; Berry 1971). Even so, it is important to recognize the difficulties inherent in Williamson's approach. Various problems are caused by deficiencies in his empirical data, particularly the poor representation of less-developed countries. Of the 24 countries in his cross-section, only Chile, Brazil, Colombia, the Philippines and India can genuinely be considered as Third World countries, with the possible additions of Greece, Spain and Puerto Rico. As Williamson clearly notes, this limitation severely qualifies the empirical support for the idea of early divergence. Similar difficulties are encountered with the time-series data; among the ten countries considered, only Brazil clearly is a less-developed country.

There are also major problems relating to the reliability and presentation of the data. The income concepts used in the paper are not identical for each country. In most cases income per capita is used, but net product at factor cost per capita, declared income per capita, median family income and personal income per family are also employed. Clearly, the use of different income concepts itself generates differences in the level of geographical concentration between the countries. In addition to this problem, the general difficulty of calculating per capita income figures in less-developed countries should be remembered. In particular, the problem of assessing the value of agricultural and subsistence production makes accurate accounting highly problematical (Elliot, 1972). Estimates of regional income also vary within particular countries according to the source. In Colombia, for example, different estimates of per capita income differentials in 1964 between the richest and the poorest departments gave figures as varied as 4 to 1 and 9 to 1 (Marabelli, 1966; Daza Roa, 1967). Variations of this magnitude clearly raise further doubts about the significance of Williamson's findings.

Difficulties in interpreting Williamson's findings also arise from use of the coefficient of variation and the adoption of arbitrary regional divisions (Metwally and Jensen, 1973). In many respects the coefficient of variation provides an acceptable measure of disparities.[1] It is sensitive to transfers at all levels in the distribution and is neutral in so far as it attaches equal weights to transfers at different levels of the distribution (Sen, 1974). In addition, Williamson's population weighted coefficient takes into account the differential importance of regions with widely different populations. The importance of this modification is indicated by the difference recorded by Williamson's measures V_w and V_{uw}. Unfortunately, however, the weighted index does not take account of variations in the size and number of administrative units on which the data are based (Parr, 1974). As a result, different levels of aggregation among the regional units can lead to important variations in the values of the indices. The V_{uw} index for Kenya in 1962, for example, ranges from 2.20 to 0.85 according to whether Nairobi is defined as a separate entity or aggregated with the neighbouring Central Province. The V_{uw} coefficient reacts still more erratically to this modification, varying from 4.23 when Nairobi is considered separately to 0.81 when it is included in the larger area. A possible cause of a high index of regional inequality, therefore, is the nature of the administrative division of the country. If the major city is included with a rural area, the level of inequality will be less marked than if it is considered alone. Indeed, we suggest that this factor is the principal explanation of the low level of regional inequality recorded for India in Williamson's data. The large rural–urban disparities in that country are hidden by the inclusion of most urban centres within administrative areas containing large rural populations. In contrast, the data for the Philippines, which show a high level of regional inequality, are based on a regional distribution which includes metropolitan Manila as a separate entity. It is probable, therefore, that the recorded difference in regional inequalities between the two countries is due as much to differences in regional units as the level of development. The point is especially important in so far as India represents the only real evidence in favour of Williamson's proposition that small disparities are characteristic of nations at low levels of development.

Variations in the number of regional divisions in different countries may also have a further effect on Williamson's findings. He includes a number of small countries with numerous regional divisions along with several large countries with limited numbers of administrative units. The survey includes Australia with 6 states, the United States with 9 regions, Puerto Rico with 76 *municipios* and Japan with 46 states. Clearly, this fact may account for the secondary finding that 'given the level of national development, the larger the geographic size of the national unit, the greater will be the degree of regional inequality' (Williamson, 1965, page 15).

Notwithstanding these difficulties, Williamson's data are consistent with his main hypothesis. Whether measured over time within a single nation or on a cross-national basis, there is a clear tendency towards convergence in developed countries. His other finding, that the early stages of development are marked by regional income divergence, however, receives only flimsy support from his evidence.

Of course, the possibility of early divergence is advanced by several development theorists, as noted below. It is also supported by evidence that other measures of economic activity, urban concentration and infrastructural provision (Friedmann, 1966; Adelman and Morris, 1973; Berry, 1961; Odell and Preston, 1973) appear to become more spatially concentrated in the process of economic growth in low income nations. For these reasons, it is appropriate to supplement Williamson's analysis with the aid of more recently published regional income data.

Regional data relating to the 1960s have been collected for fifteen less developed nations. The countries for which data are available clearly are not representative of all the nations in the Third World. Most of the countries are Latin American, which reflects their more developed economic and statistical services and the authors' greater familiarity with that region. The data also contain most of the weaknesses described for Williamson's original study. Despite these various kinds of bias, the data provide additional information about possible divergence in regional incomes at low income levels.

The available cross-national evidence, presented in Table 5.2 and Figure 5.1, shows no marked tendency for regional income differentials to increase with rising levels of per capita income. The largest differentials are found in Tanzania and Kenya where per capita incomes are among the lowest of the countries considered. On this evidence, it appears that regional disparities occur at an extremely early stage of economic development. Although it is invalid

Table 5.2 Regional inequality in selected less-developed nations

Nation	GNP per capita 1967[1]	V_w	V_{uw}	Date	Number of regions
Venezuela	880	.66	.74	1969	9
Argentina	800	.45	.59	1969	22
Spain	680	.30	.28	1971	50
Mexico	490	.65	.53	1965	32
Chile	470	.35	.60	1967	7
Peru	350	.53	.42	1961	23
Colombia	300	.24[2]	.29[2]	1964	15
Brazil	250	.60	.58	1969	21
Ghana	200	.55	.72	1960	7
Philippines	180	.64	.75	1966	11
Bolivia	170	.57	.64	1967	9
Thailand	130	.55	.50	1969	4
Kenya	120	.85[3]	.81[3]	1962	6
India	90	.17	.17	1964/5	14
Tanzania	80	.63[4]	.32[4]	1967	17

1 IBRD World Bank Atlas of per capita production and population, 1969.
2 Using the data of Daza Roa (1967), $V_w = .59$ and $V_{uw} = .61$.
3 Separating Nairobi from Central Province, $V_w = 2.20$ and $V_{uw} = 4.23$.
4 Separating Dar es Salaam from the Coastal Region, $V_w = 1.30$ and $V_{uw} = 2.08$.

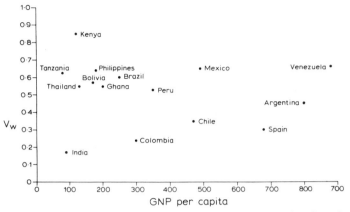

Figure 5.1 Relationship between V_w and per capita gross national product

to interpret these data in terms of 'trends' towards either convergence or divergence, several observations are appropriate.

Income differentials are smaller in several of the richer countries examined. At the same time, differences in the level of regional inequality between Tanzania, at the bottom of the income scale, and Venezuela at the top are negligible. Within the range from 80 to 880 dollars per capita, there is only slight variation in the level of regional inequality. These data, therefore, fail to lend support to Williamson's hypothesis of regional convergence with rising levels of per capita income, at least within the income range considered.

In addition, our limited data on regional inequalities over time offer no consistent support for the idea of convergence nor for divergence. Convergence occurred in India, with the V_w measure declining from 0.28 in 1955–1956 to 0.17 in 1964–1965, in Brazil between 1939 ($V_w = 0.78$) and 1969 ($V_w = 0.60$), in Spain between 1955 ($V_w = 0.37$) and 1971 ($V_w = 0.30$), in Colombia between 1951 ($V_w = 0.62$) and 1964 ($V_w = 0.59$) and in Mexico between 1950 ($V_w = 0.76$) and 1965 ($V_w = 0.65$). On the other hand, five other countries experienced regional income divergence: Chile between 1958 ($V_w = 0.27$) and 1967 ($V_w = 0.35$), Tanzania between 1963 ($V_w = 0.54$) and 1967 ($V_w = 0.63$), Argentina between 1959 ($V_w = 0.32$) and 1969 ($V_w = 0.45$), Thailand between 1960 ($V_w = 0.47$) and 1969 ($V_w = 0.55$) and the Philippines between 1948 ($V_w = 0.59$) and 1966 ($V_w = 0.64$). On the basis of these partial data, therefore, no consistent association between levels of regional income inequality and economic development can be detected.

THE INEVITABILITY OF CONVERGENCE

Although the literature on regional growth and income differentiation is vast, comparatively little is yet known about the processes likely to lead to regional income convergence. The most detailed studies of convergence have been carried out in the United States. Here convergence has been the norm since 1880 with only a brief interruption during the 1920s. According to Easterlin (1958), the main factors in this convergence are the equalization of the output structures in the different regions, the decline in the relative importance of the food industry, the growth of services in the poorer regions and the migration of labour from low income to high income regions. Interestingly enough, the movement of industry is not deemed to be of major importance.

Despite these various pressures leading towards convergence, however, the process is a slow one and has continued for more than half a century. Easterlin explains this slow convergence in the following terms. The factors that 'frequently worked against convergence were the dynamic ones: the development of new products and new techniques, resource depletion and discovery, changes in the structure of transfer costs, shifts in consumption patterns and so on ... This is not to imply, of course, that these dynamic factors necessarily worked towards divergence of regional income levels ... But clearly these factors contain no inherent bias in the direction of convergence' (Easterlin, 1958, page 325).

As a source of hope for the less-developed countries, therefore, the conclusions of this study are not wholly encouraging. While they do not rule out the possibility of convergence, they make it appear less inevitable than many writers would assume. Similarly, no firm conclusion about the likelihood of convergence can be found in the theoretical literature. A wide spectrum of views exists, ranging from belief in rapid and inevitable convergence to fear that divergence is an unavoidable result of existing economic and social systems.

According to the neo-classical economic view, the persistence of regional income differentials arises from dynamic adjustment lags due to the malfunctioning of equilibrating mechanisms (Borts and Stein, 1964; Okun and Richardson, 1961). Such differentials are regarded as a perverse result, a temporary aberration attributable to market 'imperfections' and 'institutional obstacles' which impede factor and resource movements and so prevent an efficient spatial allocation of resources. These 'imperfections' will disappear during the process of economic growth due to the unification of factor markets and the greater interdependence of regional economies. In short, market forces can be relied upon to equalize regional per capita incomes as the economy proceeds from 'under-development' through 'take-off' to 'economic maturity'.

Other economists have long questioned the effectiveness of market forces in removing regional disparities. Both Mydral (1957) and Hirschman (1958), for example, agree that poor areas are subject to strong negative forces which tend to counteract the positive 'trickle down' and 'spread' effects of economic growth. Over time, in fact, the balance of these two sets of forces will normally increase regional disparities. Powerful centripetal forces such as selective migration, the failure of entrepreneurs to perceive investment opportunities in the periphery, and the negative effects of manufacturing development on artisan industry, lead to widening disparities. In Hirschman's view, it is only the entry of government into the field of regional development which will eventually remove these disequilibria (Hirschman, 1958, page 190). On the other hand, Myrdal is more pessimistic. He believes that government action will be long delayed, and if taken at all, will tend to strengthen the existing processes towards divergence (Myrdal, 1957, page 34–35).

Myrdal's pessimistic view is developed further in the writing emerging from the less-developed nations (Frank, 1967; González, 1964–1965; Sunkel, 1969). The 'dependency' school argues that spatial inequality and the 'marginality' of the peripheral populations are inevitable consequences of the position of poorer regions in the 'development process'. These regions are linked in a satellite relationship to dynamic national centres, which in turn are linked in an external 'dependency' relationship with dynamic foreign centres. The poor peripheral regions, therefore, form the bottom rung in an exploitive system which brings development to the metropolitan economies by draining the poor areas of their economic surplus. Granted the strength of the 'dependency' relationship, the ability of governments to remove inequalities must be limited. And since incumbent political and socio-economic elites are also the beneficiaries of the 'dependency' relationship, little compensatory action can be expected from national governments.

These theories lead to different conclusions about long-term trends in regional incomes. They also attribute differential importance to the processes which affect regional income levels over time. To the neo-classicists, the changing structure of the economy will eventually lead to the reduction of spatial differences in factor prices. For the Hirschman school, it is not the unfettered operation of market forces but the intervention of government which brings about convergence. For the internal colonialists, the social and political barriers to change which exist in less-developed societies are sufficiently strong to preclude convergence.

These different views recommend that care be taken in the interpretation of regional trends. Certainly there is no reason to assume that the processes which have led to convergence in the United States and other developed countries will function automatically and effectively in the less-developed nations. Convergence may occur, but there are a number of reasons why such a process is likely to be very weak.

Firstly, it is possible that many less-developed countries may never reach the levels of development which characterize the developed countries today. To some this may seem a pessimistic conclusion but an increasingly powerful body of opinion is emerging which supports this view for a variety of reasons (Illich, 1973; Schumacher, 1973; Meadows, 1972; Commoner, 1971; Nicholson, 1970). In this event, regional inequalities will not be removed by the same process which characterized convergence in contemporary developed nations. Either political action must be taken now or inequalities are likely to remain for many generations ahead.

Secondly, hopes for ultimate convergence in regional incomes are not encouraged by a comparison between the levels of regional disparity in the less-developed nations today and those once characteristic of the developed countries during their period of maximum divergence. Williamson's Table five shows that historical values of the V_w measures in the nine developed countries during the recorded period were never greater than 0.53 in Sweden, 0.42 in Norway, 0.41 in the U.S.A., 0.38 in Italy, 0.31 in France, 0.30 in the Netherlands, 0.24 in Canada and Germany, and 0.12 in the United Kingdom. By contrast, the evidence in Table 5.2 shows that among the less-developed countries today, only Argentina, Chile, India and perhaps Colombia record figures as low as these [2]

Thirdly, regional income convergence also will depend upon government attitudes to inequality and the will to prosecute equalization policies vigorously. Hirschman (1958) assumes, optimistically, that the tolerance for regional inequality is low and suggests that a point is quickly reached where central governments are forced to implement powerful compensatory measures. However, this threshold is likely to vary between countries. *Inter alia*, the response to regional inequality will depend upon the political system in question and the strength of regional representation in national politics (Elliot, 1972; Friedmann, 1972–1973).[3] Finally, it may be observed that recent experience of regional development planning does not recommend the view that countervailing government action will have a marked impact on regional disparities (Stöhr, 1972; Gilbert, 1974).

REGIONAL INCOME EQUALIZATION AS A POLICY OBJECTIVE

Neo-classical economics notwithstanding, spatial income equalization is now widely accepted as an appropriate and desirable target for regional policy in both developing and mature economies. However, it does *not* follow that attainment of this goal will reduce personal inequalities within society or produce a more efficient spatial economic structure.

First, the actions required to achieve greater regional equality are clearly affected by the level of disparities which exist in a country. Unless the scale and causes of regional disparities are fully understood, attempts to equalize regional incomes may be aiming at an arbitrary goal (Odell, 1971). As Bowen (1970) points out with respect to personal inequalities in income, 'equality is easy to define and measure; it occurs when every income receiver gets the same sum in income. But inequality may take innumerable forms, and by any standard of measuring it must be arbitrary' (page 29). This difficulty also applies in the case of regional disparities. Greater equality may be achieved, for instance, by reducing the share of income received by people in the richest department and increasing that received by those in a middle-income department. But is the greater equality achieved by such redistribution as 'desirable' as that attained by enlarging the share of income of the people in the poorest departments? Are relative disparities the only criterion to be used or should attention also be directed towards the absolute inequalities? Clearly, these distinctions are important and can be resolved only on the basis of value judgements by the government concerned. Whatever the decision, however, it is apparent that the goal of greater equality is indeterminate unless closely qualified (Reiner, 1973).

Secondly, even if agreement can be reached on which groups of states should be the main beneficiaries of redistribution and on the period over which equalization will take place, the means of achieving equity must still be determined. Moreover, the choice of strategies to be employed itself raises important questions of equity. How, for instance, should planners decide between the age-old planning alternatives of providing employment opportunities in the backward regions or encouraging the unemployed to move to the dynamic growth areas? The per capita income of a highly populated 'downward transitional' area may be raised by encouraging out-migration. In turn, however, selective migration may reduce still further the growth opportunities of the region. This may well represent an undesirable 'solution' for the community involved, especially if, for ethnic, racial or cultural reasons, they feel themselves to be distinct from the rest of the nation. On the other hand, the alternative policy of directly creating new job opportunities may absorb scarce resources and lead indirectly to a reduction in the incomes of people in other regions. There is no simple or 'equitable' solution to this kind of problem. It is usually resolved in terms of political expedience. The fundamental point here is that the goal of greater income equality can sometimes only be achieved by inequitable means.

Thirdly, brief mention must be made of the question of equity versus efficiency in the choice of goals for regional policy. In countries following a rapid industrialization strategy, the achievement of efficiency goals may be closely related to spatial concentration. This would be the argument of regional economists such as Alonso (1969). In such a case, it is difficult to disperse economic activity too widely (even if this can be accomplished) without producing uneconomic distributions of economic infrastructure and dissipating possible economies of scale and agglomeration. One solution is some form of 'growth centre' policy, but even then it may not be viable to establish an industrial 'growth centre' in every region. Clearly, factors such as the size of country and the number of regional divisions also affect this conflict.

In certain cases, however, it is possible that development strategies may be adopted which do not pose the same kind of conflict between equity and efficiency. Rural-based economic strategies such as Tanzania's *Ujamaa* policy may embrace decentralization without dissipating unnecessarily potential economies of scale. It may also be argued that even if an industrialization policy is adopted, a measure of efficiency should be sacrificed to guarantee future equity. One reason for this approach is that those regions or groups which obtain the benefits of growth now will be most reluctant to share them with other groups or regions later. As Galbraith (1958) says, 'with increasing well-being all people become aware, sooner or later, that they have something to protect' (page 84).

Finally, in specifying the aims of regional policy, care should be taken to distinguish between the spatial and interpersonal dimensions of economic inequality. The introduction of policies which raise the average income of a poor region cannot be assumed to reduce the absolute poverty of the poorest groups within that region. Such a proposition emerges from confusing 'place poverty' with 'people poverty' and the means and ends of regional policy (Parr, 1973). In addition, it is highly questionable to claim that measures to attenuate spatial disparities will also reduce personal inequality either in the backward region or in the country as a whole.

We turn first to the question of the personal income distribution in the backward region. Progress towards the national goal of spatial income equality requires merely that *relative* income differentials diminish. For this purpose, per capita income growth in the backward region must exceed that at the national level. In short, the objective of spatial equity can be pursued by maximizing the growth rate of regional output per head relative to the national average. The speed with which income equalization occurs will depend on the weight national policymakers attach to aggregate economic efficiency and spatial equity, assuming that these goals conflict.

The central point to recognize, however, is that spatial equity and greater personal equality within the backward region may not be mutually consistent policy objectives. Indeed, diminishing relative income differentials at the national level, achieved by efficiency-orientated or output-maximizing measures, may be accompanied at the regional level by rising personal income concentration. The nature of the outcome in specific cases will be determined *inter alia* by the policy mix adopted in regional strategies to attain spatial equity. For example, the successful generation of regional growth may reduce absolute poverty by increasing the supply of employment opportunities for unskilled workers. Nevertheless, this change does not represent an unambiguous improvement in social welfare if middle and upper income groups obtain a larger and disproportionate share of these income gains, thereby aggravating the inequality of the distribution of personal income. There is obviously a possibility here for conflict between spatial income equalization and the maximization of personal welfare in the region. This can occur, assuming that the utility function is concave, since a lower level of per capita income may represent the same utility level as a higher income which is less equally distributed (Atkinson, 1970; Sen, 1974).

Statements that regional income equalization will diminish total income inequality also must be treated with a similar degree of circumspection. First, the relation or contribution of regional income differentials to total personal income inequality may be insignificant. Economic inequality is associated primarily with personal and other characteristics, including age, sex, education, occupation, etc. In Brazil, where both personal income concentration and spatial differentials are marked, income variations among regions do not contribute substantially to the observed total inequality in personal income distribution (Fishlow, 1972; Langoni, 1973). Existing personal economic inequality would persist even if full regional equalization of income were to occur. One corollary is that regional policies must attack directly the root causes of poverty, such as illiteracy, unemployment and the unequal distribution of wealth in order to modify total income inequality substantially (Fishlow, 1972).

These brief comments warn against the ambiguous and often erroneous claims at times made in support of regional income equalization. Extreme caution is needed when interpreting trends in relative income differentials in equity terms. Specifically, it is unwarranted to derive conclusions regarding changes in social welfare from evidence on movements in indices of regional equality alone. Depending on the development strategy adopted, these trends associated with many widely differing configurations of the aggregate and regional distributions of personal income. In short, movements in regional income differentials and social welfare should not be confused. A further implication is that once regional income equalization is adopted as a policy goal, policymakers must chose between alternative patterns of growth and distribution. That is, the welfare implications of regional income equalization will depend primarily on the means selected to achieve this objective. Policymakers inevitably must introduce values so that the alternative distributions can be ranked.

This discussion has noted some of the complex issues raised by regional income equalization. National and regional planning, as well as academic writing, frequently ignores these issues or offers a confusing juxtaposition of spatial and interpersonal equity goals with both national and regional efficiency objectives. For example, it is often claimed that measures to attain 'self-sustained regional growth', 'autonomous regional development' or some related efficiency goal will simultaneously bring greater equality in both the total and regional personal income distribution. However, unless the welfare equity constraints are clearly specified, these regional strategies are more properly regarded as pursuing subnational efficiency goals (Alonso, 1969). It stands repetition that the measurment and evelution of spatial and personal equality raise distinct and separate issues. The principal link between these goals resides in the choice of policy means or strategies since this will determine which groups and classes are the main beneficiaries of growth.

REGIONAL POLICIES AND INCOME CONVERGENCE IN BRAZIL

Severe regional disparities have long characterized Brazilian growth patterns and have received central government attention intermittently since the late nineteenth century (Hirschman, 1965; Robock, 1963). The proximate cause of this concern has been the periodic droughts which afflict the region's semi-arid interior, disrupting the predominantly rural economy and causing famine and human suffering on an epic scale. The initial focus of federal efforts on drought relief and water-resource utilization has gradually been superceded by programmes with wider development objectives. Concomitantly, regional planning and policy has assumed an increasingly prominent position in Brazilian development plans Dickenson (1974). The regional development agency, SUDENE, was established in 1959 and has implemented a large scale industrialization programme, known as the 34/18 scheme (Goodman and Cavalcanti, 1974; Hirschman, 1968). Although this scheme was the principal instrument of federal development policy in the Northeast during the 1960s, it was complemented significantly by federal tax transfers and ambitious transportation and power programmes (Barboza de Araujo, 1973; SUDENE, 1973). Regional policies were further diversified following the 1970 drought to embrace agricultural credit programmes and the development of rural infrastructure.

The character and execution of Brazilian regional policy since the 1964 *coup* have been heavily influenced by the ideological orientation of the military regime. This is evident not only in the failure to implement the redistributive components of the original GTDN programme but also in the identification of national income growth as the primary focus of policy concern (GTDN, 1959). The merits of free enterprise and the efficacy of market solutions are central tenets in the post-1964 Brazilian 'model'. Creation of a vigorous free market system is espoused as a major policy goal. One important corollary of this commitment is that efficiency objectives are given top priority in policy decisions. More specifically, national development strategy eschews any *direct* concern for income equality and redistributive objectives in favour of rapid economic growth and higher levels of per capita income. These aims have been pursued with some success and the Brazilian 'miracle' has been widely publicized by the news media. However, large segments of the population have secured only negligible gains in real income and the concentration of income has increased dramatically (Fishlow, 1972, 1974; Langoni, 1973).

This efficiency-orientated strategy outwardly incorporates equity-related goals, such as employment expansion, but these in practice are sought only as indirect results of rapid income growth. Improvements in the welfare of the poor and under-privileged depend on the rate at which the benefits of aggregate income growth 'trickle down'. The preference for this uncertain and random distributional process over programmes specifically designed to improve the conditions of low income groups emphasizes the preeminence of economic efficiency over social justice.

As intimated earlier, the paramount importance of efficiency considerations has strongly affected the choice and conduct of regional policies in the Northeast. The 34/18 industrialization scheme, with its overwhelming emphasis on subsidies to fixed capital formation, neatly illustrates this point. This powerful set of investment incentives does not include one measure which directly reduces total labour costs to private industrialists. This orientation has been followed for over a decade in a region distinguished by severe labour underutilization. Similarly, public expenditure policy has channelled investment resources primarily to transportation, communications, power and secondary activities. Capital investment in such social services as public health, sanitation, housing and education, which have a more direct impact on the poorer, less privileged groups in society, has been seriously neglected.

Although Brazilian plans refer variously to the reduction of regional 'disparities', 'gaps', 'dualism', 'disequilibria', etc., these terms are used interchangeably and refer solely to per capita income differentials. There is also unanimity that regional income equalization requires the adoption of measures to accelerate and sustain rapid growth. This emphasis on efficiency objectives emerges clearly from the regional development programmes elaborated by SUDENE in the 1960s. Indeed, this line can be traced to the 'reformist' programme proposed by the GTDN in the late 1950s to achieve self-sustained growth and 'autonomous regional development'.

Finally, the current national plan (*II Plano Nacional de Desenvolvimento 1975–1979*) will ensure the continuity of this approach. 'The orientation of macro-economic policy in the Northeast is to ensure accelerated growth and give it a self-sustained character ...' (page 54). Specifically, the Northeast is set the formidable task of achieving annual growth rates exceeding the national target rate of 10 percent '... in order to reduce the existing gap' (page 53–54). It is suggested that this strategy will generate employment opportunities and rising real incomes for all sections of Northeastern society and also promote income redistribution. However, apart from the implicit assumption of 'trickle down' effects, it is nowhere demonstrated how these aims will be secured. Experience of the 1960s and early 1970s certainly does not encourage a sanguine view regarding the equitable distribution of the benefits of future income expansion.

Interregional income disparities

Unequivocal evidence on the direction of trends in regional income differentials in Brazil is not readily available, although the post-war data can be interpreted to support the convergence hypothesis (Almeida Andrade, 1974; Gauthier and Semple, 1974). (See Table 5.3.) The question of interpretation and judgement is raised advisedly since acceptance of this proposition requires willingness to ascribe a long-term trend to small changes in data of dubious validity. As the ratio for 1939 demonstrates, any conclusion on this matter is extremely sensitive to the choice of base period. Furthermore, exceptional

Table 5.3 Changes in regional shares and regional product per head in Brazil, 1939–1968

Period	Regional Shares in Net Domestic Product[1]			Regional Net Product Per Head in Relation to the National Average[1]		
	North	North–East	Centre–South	North	North–East	Centre–South
1939	2.7	16.7	80.6	.74	.48	1.31
1947–1953	1.9	14.2	83.9	.52	.40	1.36
1954–1961	2.1	13.6	84.3	.61	.42	1.32
1962–1968	2.0	14.8	83.2	.53	.48	1.27
1947–1968	2.0	14.2	83.8	.54	.47	1.28

Source: Fundação Getulio Vargas; Fundação Instituto Brasileiro de Geografia e Estatística.

Notes:
1. Calculations based on estimates of net domestic product at factor cost in current prices. The states included in each region are shown in Figure 5.2.

natural or climatic events may occur, as in the period 1947–1953 when the expansion of Northeastern output was severely restricted by the prolonged drought of 1951–1953. These circumstantial factors are reinforced by statistical deficiencies which complicate regional income comparisons. Despite regional differences in the cost of living and consumption patterns, it is customary to use price deflators elaborated for the higher income and more diversified economy of the Centre–South (Figure 5.2). Regional differences in the aggregate and sectoral composition of output also are neglected and implicit price deflators drawn from national income data typically are used in the calculation of real product series. The possibility of serious distortion is illustrated perfectly by the fact that the official statistical agencies present widely differing estimates of the sectoral distribution of regional product. According to SUDENE, secondary activities accounted for 22 percent of regional output in 1965–1968 whereas the Getulio Vargas Foundation gives an estimated share of 10 percent (C. Cavalcanti, 1972).[4] Furthermore, as their respective estimates of real product growth in the years 1960–1968 are 6.4 and 4.2 percent, the trend in

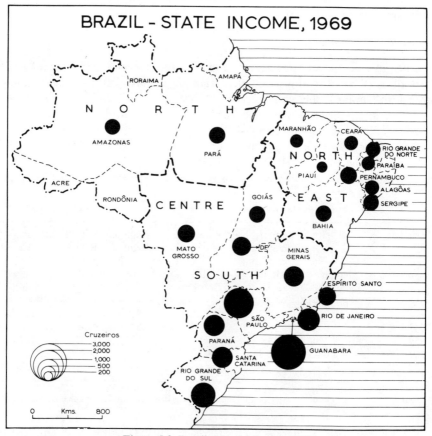

Figure 5.2 Brazil—State income, 1969

relative income disparities will depend significantly upon which series is selected. The essentially arbitrary nature of this choice reveals the dangers of attributing long-term movements of convergence or divergence in per capita income differentials to minor changes which fall well within the range of statistical error.

The evidence to hand is thus inconclusive, which recommends the postulate that regional income differentials have remained fairly stable (Gilbert, 1974).[5] Data limitations apart, it appears that positive net out-migration has offset the negative effects of lagging output growth on per capita income in the northeast (Graham, 1970; da Mata, 1973). The divergence of northeastern and national rates of population growth was a potent force for regional income equalization in the 1950s, with respective annual rates of 2.2 and 3.1 percent. This difference has since narrowed and population census data for the period 1960–1970 give annual average growth rates of 2.5 percent for the northeast and 2.9 percent for Brazil. With the weakening of this compensatory mechanism, movements in per capita income differentials will depend more on the Northeast's relative growth performance. Whether this challenge can be met successfully raises issues beyond the scope of this paper. However, preliminary real product data for the period 1967–1973, when the Brazilian 'miracle' was at its peak, indicate that the Northeast achieved an annual growth rate of 7 percent, significantly below the national average rate of 10 percent.

Income inequality within the Northeast

It is appropriate now to assess the impact of regional strategy on equity-related measures such as income distribution, unemployment and poverty. Income inequality increased sharply during the 1960s, the Gini coefficient rising from 0.49 to 0.56 and the share of the poorer 50 percent in regional income falling from 18.6 to 16.3 percent (Langoni, 1973). The extreme inequality of the regional income distribution is strikingly shown by the rise in the share of the richest 10 percent to 47.1 percent, virtually three times that of the lower half of the region's population.

The clear trend towards increasing income concentration in the Northeast is particularly evident in urban areas, the main targets and beneficiaries of the region's industrialization programme. Gini coefficients estimated from census data show rising inequality in the urban sector between 1960 and 1970, especially in secondary activities, and an improvement in equality in agriculture (Goodman and Cavalcanti, 1974; Hume, 1972). These trends have been attributed to the migration of unpaid family workers and the rural unemployed into low-productivity urban employment and the diffusion of more capital-intensive techniques into the region's industrial sector (R. Cavalcanti, 1970). Although Langoni (1973) does not present disaggregated sectoral data when discussing trends in regional income distribution for the 1960s, he does conclude that '... the high index of concentration observed for the Northeast (Gini of 0.57) is wholly explained by the behaviour of the urban sector (Gini of

0.60) since the primary sector presents a substantially lower degree of inequality (Gini of 0.37) ...' (Langoni, 1973, page 166).

The weight of evidence for the 1960s points unequivocally towards a pronounced deterioration in the equality of income distribution. This trend is particularly evident in the urban sector. The present situation is that the share in urban income of the poorest 30 percent of the urban population is 6 percent, whereas the richest 30 percent, obtains approximately 71 percent. The Northeast thus provides a vivid illustration of how equity considerations can be neglected when regional income equalization is formulated as a subnational efficiency objective. Of course, in this respect, regional policy represents a microcosm of national growth strategy. Nevertheless, the Northeast now exhibits the most highly concentrated income distribution among the country's six major regions and remains the area of greatest absolute poverty (Langoni, 1973).

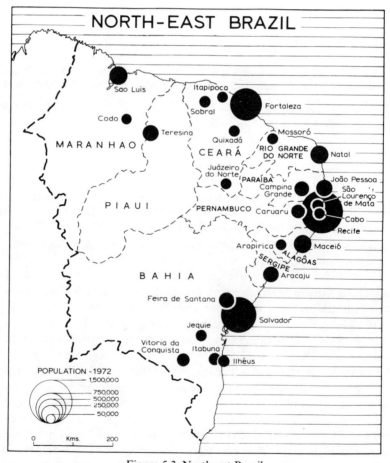

Figure 5.3 Northeast Brazil

This evidence of increasing economic inequality severely qualifies any claims that the 1960s witnessed a period of development, irrespective of progress towards regional income equalization.

It is extremely difficult to complement this evidence with time series of employment and real income data for different socio-economic groups. The burden of our discussion thus will fall on data for the early 1970s in an attempt to depict the gravity of the region's development problems. The exception is provided by the family budget surveys undertaken by the Bank of the Northeast

Table 5.4 Changes in family income distribution and real income per head in sleected northeastern cities

Proportion of Urban Population	Recife	Salvador	Fortaleza	Natal	Maceió	João Pessoa	São Luis	Campina Grande

A. Family Income Distribution (percent of Income)

Survey I: Date	Oct. 1960	1962	1962	Nov. 1964	Apr. 1964	Nov. 1964	Sep. 1963	May 1962
Lower two-thirds	37.7	34.3	40.3	35.4	32.8	34.9	40.3	36.4
Upper third	62.3	65.7	59.7	64.6	67.2	65.1	59.7	63.6
Survey II: Date	Mar. 1967	Aug 1966	Jul. 1966	Jul. 1966	Mar. 1968	Jul. 1967	Feb. 1967	Jul. 1967
Lower two-thirds	28.3	32.2	34.4	34.0	29.9	29.9	32.4	25.6
Upper third	71.1	67.8	65.6	66.0	70.1	70.1	67.6	74.4

B. Monthly Real Income per head.[1]

Survey I								
Lower two-thirds	37.8	44.7	43.3	32.2	31.9	32.4	41.8	23.8
Upper third	125.1	172.0	128.1	117.7	131.0	121.0	124.3	83.3
Total Population	*66.8*	*87.1*	*71.5*	*60.6*	*64.9*	*61.9*	*69.3*	*43.6*
Survey II								
Lower two-thirds	36.8	54.6	37.8	31.3	36.4	44.6	46.5	30.9
Upper third	186.8	230.4	144.6	123.2	170.2	215.2	194.4	179.5
Total Population	*86.8*	*113.1*	*73.6*	*62.1*	*81.1*	*101.4*	*95.8*	*80.4*

Source: Banco do Nordeste do Brazil/Superintendéncia do Desenvolvimento do Nordeste.
1. In new *cruzeiros* at April, 1969 prices.

in the region's eight largest cities during the 1960s (BNB–ETENE, 1969). (See Figure 5.3.) These surveys strongly support the thesis that urban income inequality increased and also reveal the restricted gains made by groups at the lower end of the distribution (Table 5.4). Indeed, in Recife, Fortaleza and Natal, the declining share of the lower two-thirds of the urban population is accompanied by an *absolute* fall in average real income. This depressing point is reinforced by an additional finding of the BNB surveys that real income per head of the poorest quintile of the urban population diminished in all the cities considered here except São Luis.[6]

A recent case study of family incomes in Recife furnishes additional evidence on this point. Recife is the region's principal metropolitan area and is frequently identified as a major beneficiary of the current regional policy, particularly the 34/18 industrialization programme. The spatial distribution of investment under this scheme is highly concentrated and Recife alone has attracted roughly 25 percent of total project outlays. With this pre-eminent position and evidence that the city's population growth is diminishing, one would reasonably expect to find a measurable rise in per capita real income in the 1960s. However, the results of family budget surveys for 1960–1962 and 1967–1968 completely confound this view. Even on the most favourable interpretation, the annual growth rate of real income per head in Recife does not exceed 1.5 percent from 1960 to mid-1967, rising from U.S. $277 to U.S. $308 at June 1971 prices. The level of average monthly income to support an average size family of six people fluctuates between U.S. $141 in 1960, U.S. $153 in 1961–1962 and 1967 and U.S. $140 in early 1968. The stagnation of *average* real family income, allied to extreme urban income inequality, implies that the real income of families at the lower end of this distribution almost certainly declined. As the author observes, '... this denotes the existence of population groups living in conditions possibly below the subsistence level' (C. Cavalcanti, 1972a, page 91–92). In less guarded terms, these surveys provide *prima facie* evidence of starvation and severe malnutrition in the region's 'capital' city.

The results of these family budget studies are particularly significant in the absence of real income trends for separate population deciles of the income distribution during the 1960s. Thus Langoni's study indicates merely that *average* urban real income per head increased by 53.3 percent between 1960 and 1970, rising at an annual rate of 4.6 percent. Yet, in each of the three subregions of the Northeast, this average exceeded the income of workers in the first seven deciles of the urban income distribution, in 1970. Such inequality, together with the budget survey data, indicate that the lower income groups had a meagre share in the benefits of this expansion. Absolute real income changes for the poorer segments of the urban community appear to have been extremely modest, if indeed at all positive, and their relative income position deteriorated. These findings further underline the neglect of equity considerations in recent regional policy, which is implemented in the name of interregional income equalization.

Poverty indicators in the Northeast

Recent urban employment and earnings data again suggest that the effects of recent growth have not been diffused widely throughout the urban community. This is revealed in part by continued high levels of urban under-employment, involving an estimated 20–25 percent of the urban labour force (Pellerin, 1971; Hume, 1972). However, this phenomenon is part of the wider, more prevalent condition of low income employment and urban poverty. This is confirmed by the 1970 Census, which indicates that 581 thousand workers received under 50 *cruzeiros* or U.S. $11 per month, and 1.1. million, 36 percent of the urban labour force, earned below 100 *cruzeiros* per month. This figure is 20 percent below the legal monthly minimum wage for the Northeast, which is considered officially as the lower limit required to sustain a minimum acceptable standard of living.[7] PNAD household survey data for 1972 greatly strengthen this point since 69 percent of the Northeast's employed urban population received only one minimum salary or less. It appears entirely appropriate to describe the vast bulk of the urban labour force as the 'working poor'. That is, workers with jobs, often working long, exhausting hours, whose earnings are at or below the level established officially as the 'subsistence income' or poverty line.

The endemic nature of acute urban poverty is illustrated by the size of two lower income categories *within* the 'working poor'. The first contingent earning from one-quarter to one-half of the minimum salary includes 754 thousand workers or 18 percent of the urban labour force. The poorest group, with extraordinarily low earnings of under one-quarter of the official minimum wage covers a further 902 thousand workers, fully 22 percent of the urban active population. The size of the 'working poor' engaged in low-wage employment and the large 'hard core' component of 1.6 million workers with desperately low earnings suggests the tenuous impact of recent growth on urban labour markets. The 'hard core' identified here, which embraces 40 percent of the urban labour force, can participate only marginally in the monetary economy. Indeed, the magnitude of the 'working poor', some 3 million strong, cruelly exposes claims that the urban proletariat can be included among the major beneficiaries of the income growth generated by regional industrialization and the Brazilian 'miracle'. Moreover, when this expansion was at its peak in 1968–1973, 89 percent of the *increase* in salaried employment in the Northeast occurred in earnings classes of one regional minimum salary or less (Salm, 1974).

A similar profile of urban poverty also emerges from special tabulations of 1970 Census returns of *family* incomes in twenty-one Northeastern cities (Goodman, 1974).[8] In the case of the region's nine state capitals, roughly one-third of all families subsist on monthly incomes of 150 *cruzeiros* or less, roughly U.S. $33 at 1970 prices. A doubling of this extremely low income level is sufficient to embrace almost two-thirds of the households in these centres.[9] Higher ratios, varying from one-half to two-thirds, of households with monthly incomes under 151 *cruzeiros* are found in the region's smaller secondary cities. Several important interior centres such as Juazeiro do Norte, Sobral, Mossoro and Caruaru present truly alarming concentrations of households in earnings classes below 101 *cruzeiros*, markedly below the official poverty threshold. The conclusion that broad segments of the urban community exist in conditions of extreme deprivation is inescapable.

On the evidence available, there is no case for the view that the 'trickle down' effects of recent growth generated by national and regional policies have been sufficiently strong and widespread to have brought significant improvements in the living standards of the urban masses in the Northeast. This conclusion applies particularly to the 'working poor' employed in low wage occupations a category which embraces two-thirds of the urban labour force. In addition, the position of workers and urban families appreciably below the official poverty line probably deteriorated in absolute terms. Although progress towards aggregate and regional income targets has occurred, the unbalanced distributional impact of recent programmes is firmly established. The trend towards greater income concentration during the 1960s exposes the glaring deficiencies of these growth-orientated policies in promoting greater social welfare. In short, the virtually exclusive focus of regional strategy on efficiency objectives has been socially divisive, exacerbating the socio-economic stratification of Northeastern society.

Although data on income distribution, real incomes and employment reveal the continued gravity of northeastern development problems, the degree of absolute poverty and its implications tends to be obscured. Several isolated references can serve to remind us of the critical importance of this absolute dimension. A recent study concludes that the nutritional position in the Northeast, which was already critical by U.N. standards in the early 1960s, deteriorated further in the course of the decade (IBRD, 1974). Nutritional deficiencies also account for the high rates of child mortality, which in the capital cities of the states of Piauí, Ceará, Sergipe and Bahia average 108.4 per 1000 live births. The average for the whole of Brazil is 79 child deaths. Gastro-enteritis, other diarrhoeic diseases and pneumonia '... together account for over 60 percent of child deaths ...' and these diseases, '... as well as the other causes of child mortality, are aggravated, and in most cases made lethal by malnutrition ...' (IBRD, 1974). It is widely known that nutritional deficiencies, particularly if suffered at the early age of 0–4 years, adversely affect normal physical and mental growth (Scrimshaw and Gordon, 1968).

In addition to malnutrition, other observers have drawn attention to the critical role of substandard housing and primitive sanitary facilities in the transmission of debilitating diseases among low income groups. These factors are cited to explain the widespread incidence of esquistossimiasis in the densely populated shanty towns or *mocambos* of Recife (Coutinho, 1974). It would be easy to extend this list further but the implications of absolute poverty are conveyed most vividly by differences in life expectancy rates. Estimates for Brazil since 1930 indicate that life expectancy at birth has been consistently lowest in the Northeast's central and most populous region, formed by the states of Ceará, Rio Grande do Norte, Paraíba. Pernambuco and Alagoas.[10] Life expectancy in this subregion in the period 1960–1970 was 44 years whereas the national average was 56 years. Life expectancy at birth in other states was as follows: São Paulo (63 years), Paraná (62), Rio Grande do Sul and Santa Catarian (68). 'It can be affirmed that the central region of the Northeast has one of the highest levels of mortality in the world' (Carvalho, 1974, page 15). As the author emphasizes, life expectancy at birth in India was 50 years in the 1960s and he concludes that 'perhaps no other country presents regional differences in mortality as great as those in Brazil'. This new evidence of regional disparities again supports the case for measures which improve the conditions of low income groups directly.

We close this section with a plea for regional policies which are specifically concerned with the causes of low productivity and low incomes. After all, regional income equalization policies usually are introduced initially because of the concentration of *absolute* poverty in certain backward areas. The elimination of such poverty remains a major justification for regional programmes but, in practice, it is often neglected in favour of maximizing output growth. The experience of regional policy in the Brazilian Northeast provides an outstanding illustration of the shortcomings of this latter approach. The evaluation of this programme since the 1960s in terms of higher real earnings for all income groups, unemployment and poverty relief and greater equality is profoundly disappointing. In conclusion, recent programmes do not attack absolute poverty directly in the attempt to disrupt the cycle of inequality and its transmission from one generation to another. That recent policies in the Northeast have shirked this challenge by design is the true indication of their inadequacy as instruments of regional *development*.

CONCLUSION

Williamson's proposition that regional incomes first diverge and then converge as per capita incomes rise does not meet with strong support from the available evidence. The proposition therefore should be regarded with serious reservation. As the author himself notes, the data he used are unreliable, his measures of inequality sensitive to variations in the size of regional unit, and his sample of nations unrepresentative of the Third World. Recourse to more recent data for other less-developed nations to supplement his international cross-section, do not provide support either for convergence or for divergence in regional incomes.

Many academics and planners accept the inevitability of convergence even though the structural features of less-developed countries raise serious doubts about the theoretical and empirical foundations of this view. Given the difficulties facing development in these countries it is possible that per capita incomes may not rise to the high levels where regional income convergence allegedly occurs. More fundamental still is that regional disparities are unlikely to diminish unless national governments adopt strong regional development programmes. Without very determined countervailing efforts, wide regional income differences are likely to continue.

The criterion of regional income equalization should be used with care. It is not difficult to conceive ways in which regional income convergence may occur without leading either to rapid national growth or to an improvement in the situation of the poor. Specifically, regional income convergence may be associated with negligible gains in (or sometimes even a lowering of) real incomes of the poorest groups in society and with a worsening of the size distribution of incomes within the poorest regions.

The experience of Northeast Brazil illustrates the difficulties which face regional and national planners in their choice of objectives. Despite the rapid growth of the national economy, the introduction of several ingenious regional measures, and achievement of a measure of regional income convergence, the benefits to most Northeasterners are few. In urban areas, the real income of the poorest groups has declined, personal income inequalities within the region have widened over time, and measures of social deprivation, such as infant mortality rates, remain appallingly high.

To be effective in helping in the poor of the less-developed areas, regional programmes must contain measures aimed directly at increasing their access to gainful employment opportunities, extending the coverage of social services and reducing inequality. Regional programmes which are efficiency oriented and rely wholly on 'trickle down' effects to help the poor are unlikely to prove effective.

Acknowledgements

We should like to thank Chris Dixon, Charles Elliot, Arthur Morris, John Naylon, Michael Safier and the Bank of Bilbao for help with the gathering of income data, and Lauchlin Currie for making detailed comments on the text.

NOTES

1. Williamson (1965) uses two main measures of inequality (V_w and V_{uw}) both based on the coefficient of variation.

$$V_w = \sqrt{\frac{\sum_i (y_i - \bar{y})^2 \cdot (n_i/n)}{\bar{y}}}$$

where n_i = population in region i,
   n = national population,
   y_i = income per capita of region i,
   and $\bar{y}$ = national income per capita

$$V_{uw} = \sqrt{\frac{\sum_i (y_i - \bar{y})^2 / N}{\bar{y}}}$$

where N = number of regions.

2. This evidence is in some contrast to Kuznets's findings about the distribution of personal income. 'The pattern of the size distribution of income characterizing under-

developed countries today is not too different from that observed in the presently developing countries in the 1920s and 1930s, or at the beginning of the century' (Kuznets, 1963, page 68).
3. Friedmann (1972–1973) distinguishes five sets of élite relationships which are likely to affect the attitudes of national governments to regional problems. He labels these relationships cooptation, accommodation, open hostility, regional 'protectorates', and federative solutions.
4. Manufacturing, civil construction and public utilities are the activities which make up the secondary sector.
5. Estimates based on Demographic Census returns of personal pre-tax income of individuals remunerated in monetary form reinforce this conclusion. These data indicate that *relative* per capita income in the northeast declined from 56.8 to 55.8 per cent in the period 1960–1970 (Langoni, 1973).
6. These surveys embrace the years 1964–1967 when the military government implemented a major price stabilization programme. It should be recognized, however, that nominal wage restraint was an integral, not incidental, element of this policy and real minimum wages deliberately were permitted to decline. This wages policy was consistent with the regime's ideological commitment to the free enterprise system and its aim to pursue a growth strategy in which private capital accumulation would play a leading role.
7. For purposes of this discussion, the legal minimum wage will be interpreted as an official poverty threshold or poverty line. In Brazil, this wage frequently is treated analogously with the concept of a 'subsistence income', which is needed to acquire the bare necessities of life.
8. The twenty-one urban *municipios* covered in this survey had 50,000 inhabitants or over in 1970.
9. The most recent PNAD earnings for the Northeast in the last quarter of 1972 show that 22 per cent of *all* urban 'consumer units' received one-half or less of the minimum salary, 26.6 per cent from one half to one and 25.1 percent between one and two minimum salaries. A 'consumer unit' is defined as those family members which are subject to and derive benefit from the same domestic budget. See *FIBGE–PNAD: Pesquisa de Rendimentos–PNAD*–2, 4° *Trimestre de 1972*, Rio de Janeiro, 1974.
10. This subregion had 53.5 percent of the Northeast's total population in 1970.

REFERENCES

Adelman, I. and Morris, C. T. (1973) *Economic growth and social equity in developing countries* (Stanford, Stanford U.P.).

Almeida Andrade, T. (1974) 'Regional inequality in Brazil' in Thoman R. S. (Ed.) (1974) 337–360.

Alonso, W. (1969) 'Urban and regional imbalances in economic development', *Economic Development and Cultural Change*, **17**, 1–14.

Atkinson, A. B. (1970) 'On the measurement of inequality', *Journal of Economic Theory*, **2**, 244–263.

Baer, W. (1964) 'Regional inequality and economic growth in Brazil', *Economic Development and Cultural Change*, **12**, 268–285.

Banco do Nordeste do Brasil-ETENE (1969) *Distribuicão e niveis da renda familiar no Nordeste Urbano* (Fortaleza).

Barboza de Araujo, A., et al. (1973) *Transferências de impostos dos estados e municipíos* (Rio de Janeiro, IPEA–INPES, Relatório de Pesquisa, No. 16).

Berry, B. J. L. (1961) 'City-size distributions and economic development', *Economic Development and Cultural Change*, **9**, 573–588.

Berry, B. J. L. (1973) 'City size and economic development: conceptual synthesis and policy problems, with special reference to South and Southeast Asia', in Jacobson, L. and Prakash, V. (Eds.) (1971) *Urbanization and National Development* (Beverly Hills, Sage Publications).
Borts, G. H. and Stein, J. L. (1964) *Economic growth in a free market* (New York, Columbia U.P.)
Bowen, I. (1970) *Acceptable inequalities: an essay on the distribution of income* (London, Allen and Unwin).
Bugnicourt, J. (1970 and 1971) 'Conjonction, correlations et contiguité des disparités régionales en Afrique', *Cultures et Développement*, **II**, 607–626 and **III**, 67–80.
Carvalho, J. A. M. (1974) Evolucão Demográfica do Nordeste Brasileiro Comparada com a Evolução Demográfica do Brasil, 1940–1970. (mimeo).
Cavalcanti, C. (1972a) 'A renda familiar e por habitante na cidade do Recife', *Pesquisa e Planejamento Econômico*, **2**, 81–104.
Cavalcanti, C. (1972b) 'Uma Avaliação das Estimativas de Renda e Produto do Brasil', *Pesquisa e Planejamento Econômico*, **2**, 381–398.
Cavalcanti, R. (1970) 'Desenvolvimento industrial e distribuicão de renda: a Experiência Brasileira', *Revista da Secretaria da Fazenda*, 1st Semester, **1**, 4–20. Recife.
Commoner, B. (1971) *The closing circle: confronting the environmental crisis* (London: Cape.).
Coutinho, A. B. et al (1974) 'Ecologia Urbana dos caramujos do gênero biomphalaria', *Revista Pernambucana de Desenvolvimento*, **1**, 29–44.
Daza Roa, A. (1967) 'La repartición de los ingresos', *Revista del Banco de la República*, **40**, 880–889.
Dickenson, J. P. (1974) 'The impact of government policy on regional inequalities in Brazil', in Thoman, R. S. (ed.) (1974), 297–320.
Easterlin, R. A. (1958) 'Long-term regional income changes: some suggested factors', *Papers and Proceedings of the Regional Science Association*, **4**, 313–325.
Elliot, C. (1972) 'Income distribution and social stratification: some notes on theory and practice', in Baster, N. (ed.) (1972) *Measuring development: the role and adequacy of development indicators* (London: Cass), 37–56.
Fishlow, A. (1972) 'Brazilian size distribution of income', *American Economic Review*, **62**, 391–402.
Fishlow, A. (1974) Brazilian income size distribution–another look (mimeo).
Frank, A. G. (1967) *Capitalism and underdevelopment in Latin America* (New York: Monthly Review Press).
Friedmann, J. (1966) *Regional development policy: a case study of Venezuela* (Cambridge Mass: M. I. T. Press).
Friedmann, J. (1972–73) 'The spatial organisation of power in the development of urban systems', *Development and Change*, **IV**, 12–50.
Furtado, C. (1963) *Economic growth of Brazil. A survey from colonial to modern times*, (Berkeley: University of California Press).
Furtado, C. (1973) 'The Post-1964 Brazilian "Model" of development', *Studies in Comparative International Development*, **VIII**, 115–127.
Galbraith, K. (1958) *The Affluent Society* (London: Hamilton).
Gauthier, H. L. & Semple R. K. (1974) 'Trends in regional inequalities in the Brazilian Economy, 1947–66', in Thoman, R. S. (ed.) (1974), 249–266.
Gilbert, A. G. (1974) *Latin American development: a geographical perspective* (Harmondsworth: Penguin).
González Casanova, P. (1964–65) 'Internal colonialism and national development', *Studies in Comparative International Development*, (**14**), 27–37.
Goodman, D. E. (1972) 'Industrial development in the Brazilian Northeast: an interim assessment of the tax credit scheme of Article 34/18', in R. J. A. Roett (ed.) (1972), 231–274.

Goodman, D. E. (1974) The Brazilian economic "miracle" and urban labour markets: a regional perspective (mimeo.)
Goodman, D. E. and Cavalcanti, R. (1974) *Incentivos à industrialização e desenvolvimento do Nordeste* (Rio de Janeiro, IPEA–INPES, Relatorio do Pesquisa, 20.).
GTDN—Grupo De Trabalho para o Desenvolvimento do Nordeste (1959) *Uma política de desenvolvimento econômico para o Nordeste* (Rio, Presidencia da Republica).
Graham, D. H. (1970) 'Divergent and convergent regional economic growth and internal migration in Brazil, 1940–60', *Economic Development and Cultural Change*, **18**, 362–382.
Hirschman, A. O. (1958) *The strategy of economic development* (New Haven: Yale University Press).
Hirschman, A. O. (1965) *Journeys toward progress: studies of economic policy making in Latin America* (New York: Anchor Books).
Hirschman, A. O. (1968) 'Industrial development in the Brazilian northeast and the tax credit scheme of article 34/18', *Journal of Development Studies*, **5**, 1–28.
Hume, I. M. (1972) *Problems and prospects in employment and incomes in Northeast Brazil* (Washington IBRD).
Illich, I. (1973) *Tools for conviviality* (London: Calder and Boyars).
IBRD (1974) *Northeast Brazil: food production and nutrition project* (Washington, D.C.: Draft Report).
International Labour Office (1970) *Towards full employment* (Geneva).
International Labour Office (1972) *Employment, incomes and equality* (Geneva).
Johnson, H. G. (1958) 'Planning and the market in economic development', *Pakistan Economic Journal*, **8**, 44–55.
Keeble, D. E. (1967) 'Models of economic development', in Chorley, R. J. and Haggett, P. (eds.) *Models in geography* (London: Methuen), 243–302.
Kuznets, S. (1955) 'Economic growth and income inequality', *American Economic Review*, **XLV**, 1–28.
Kuznets, S. (1963) 'Quantitative aspects of the economic growth of nations, VIII Distribution of income by size', *Economic Development and Cultural Change*, Part II.
Langoni, C. G. (1973) *Distribuição da renda e desenvolvimento econômico do Brasil* (Rio de Janeiro: Editora Expressao e Cultura).
Lasuén, J. R. (1974) 'National and urban development', in Banco Nacional da Habitação (1974) *Symposium on urban development* (Rio de Janeiro), 89–111.
Marabelli, F. (1966) Tentativa de distribución del producto bruto interno de Colombia por secciones administrativas del pais, 1964 (mimeo.)
Mata, M. da, et. al. (1973) *Migrações internas no Brasil* (Rio de Janeiro, IPEA–INPES, Relatório de Pesquisa No. 19).
Meadows, D. H. (et al) (1972) *The limits to growth* (London: Earth Island).
Metwally, M. M. & Jensen, R. C. (1973) 'A note on the measurement of regional income dispersion', *Economic Development and Cultural Change*, **22**, 135–136.
Myrdal, G. (1957) *Economic theory and underdeveloped regions* (London: Gerald Duckworth).
Nicholson, M. (1970) *The environmental revolution: a guide for the new masters of the world* (London: Hodder and Stoughton).
Odell, P. R. (1971) 'A European view on regional development and planning in Latin America', *International Review of Community Development*, 25–26, 3–32.
Odell, P. R. & Preston, D. A. (1973) *Economies and societies in Latin America: a geographical interpretation* (London and New York: John Wiley)
Okun, B. & Richardson, R. W. (1961) 'Regional income inequality and internal population migration', *Economic Development and Cultural Change*, **9**, 128–143.
Oshima, H. T. (1970) 'Income inequality and economic growth—the post-war experience of Asian countries', *Malayan Economic Review*, **XV**, 7–41.
Parr, J. B. (1974) 'Welfare differences within a nation: a comment', *Papers and Proceedings of the Regional Science Association*, **32**, 83–91.

Pellerin, G. (1971) *Oferta e demanda de mao-de-obra no Nordeste* (Recife, SUDENE).
PNAD (1972) *Pesquisa nacional por amostra de domicilios*.
Reiner, T. A. (1974) 'Welfare differences within a nation', *Papers and Proceedings of the Regional Science Association*, **32**, 65–82.
Richardson H. W. (1973) *The economics of urban size* (Farnborough: Saxon House).
Robock, S. H. (1963) *Brazil's developing Northeast: a study of regional planning and foreign aid* (Washington: The Brookings Institution).
Roett, R. J. A. (Ed.) (1972) *Brazil in the sixties* (Nashville: Vanderbilt University Press).
Schumacher, E. F. (1973) *Small is beautiful: a study of economics as if people mattered* (London: Blond and Briggs).
Salm, C. (1974) 'Evolução do mercado de trabalho, 1969–72', *Estudos CEBRAP*, **8**, 105–119.
Scrimshaw, N. S. and Gordon, J. E. (Eds.) (1968) *Malnutrition, learning and behaviour* (Cambridge: M. I. T. Press).
Sen, A. K. (1974) *On economic inequality* (Oxford: The Clarendon Press).
Stöhr, W. (1972) 'The state of the art: regional development programmes in Latin America at the end of the 1960s', *Latin American Urban Research*, **2**, 241–257.
SUDENE (1973) *Formação bruta de capital fixo do setor público no Nordeste* (Recife).
Sundrum, R. M. (1972) 'The distribution of incomes in the ECAFE region: causal factors and remedial policies', *Economic Bulletin for Asia and the Far East*, **XXIII**, 9–19.
Sunkel, O. (1969) 'National development policy and external dependence in Latin America', *Journal of Development Studies*, **6**, 23–48.
Thoman, R. S. (Ed.) (1974) *Proceedings of the Commission on Regional Aspects of Development of the International Geographical Union, Volume, I. Methodology and case studies* (Toronto: Allister).
UNECAFE (1971) 'Social justice, employment and income distribution in the ECAFE region', *Economic Survey of Asia and the Far East* 1971, 7–74.
UNECLA (1971) *Income Distribution of Latin America* (New York).
UNRISD (1971–75) *Reports on Regional Development* (Paris and The Hague: Mouton).
Utría, R. D. (1972) 'Regional structure and Latin American development', *Latin American Urban Research*, **2**, 61–84.
Vapnarsky, C. A. (1969) 'On rank-size distributions of cities: an ecological approach', *Economic Development and Cultural Change*, **17**, 584–595.
Williamson, J. G. (1965) 'Regional inequality and the process of national development: a description of the patterns', *Economic Development and Cultural Change*, **13**, 3–43.

Sources of regional data

Argentina 1959 Gross Regional Product—UNECLA (1968) *Desarrollo económico y la distribución del ingreso en la Argentina* (New York), Pages 84 and 234–235.
1969 Gross Regional Product—Argentina (1974) *Plan trienal de reconstrucción nacional* (Buenos Aires).
Bolivia 1967 Gross Internal Product—*Ministerio de Planeamiento y Coordinación* quoted in Boisier, S. (1972) *Polos de desarrollo: hipotesis y políticas—estudio de Bolivia, Chile y Peru* (Geneva: UNRISD).
Brazil 1939–196J Net Regional Product—Fundação Getúlio Vargas and Instituto Brasileiro d. Geografia e Estatística. See also Almeida Andrade, T. (1974).
Chile 1958 Gross Internal Product—corporacion del Fomento de la Produccion, quoted in Inst .uto de Organizacion y Administración, Facultad de Ciencias Económicas, Universidad de Chile (1962) *Estudio de recursos human os de nivel universitario en Chile* (Santiago) Part 1, pages 53 and 228.
1967 Gross Internal Product—UNECLA (1971), 'Some regional development problems in Latin America linked to metropolitanization, *Economic Bulletin for Latin America*, **16**, 57–90. Page 65.

Colombia 1951, 1964 Gross Internal Product—Daza Roa, A. (1967). Gross Internal Product—Marabelli, F. (1966). Tentativa de distribución del producto bruto interno de Colombia porsecciones administrativas del país, 1964 (mimeo.) Bogotá.

Ghana 1960 Gross Value Added—Birmingham, W., Neustradt, I and Omaboe, E. N. (1966) *A study of contemporary Ghana: Volume 1, The Economy of Ghana* (London: Allen and Unwin).

India 1955/6, 1964/5 National Income—Williamson, J. G. (1965) Page 55 Net Domestic Product (excluding defence)—Tiwari, S. G. (1971), 'Regional accounting in India', *The Review of Income and Wealth*, **17**, 103–118, Page 110.

Kenya 1962 Gross Domestic Product—Republic of Kenya (1969) *Kenya Development Plan 1964–70* (Nairobi). Ominde, S. H. and Odingo, R. S. (1974), 'Demographic aspects of regional inequalities in Kenya in Thoman, R. S. (Ed.) (1974), 555–604.

Mexico 1940, 1950, 1960 Gross Internal Product—Unikel, L. & Victoria, E. (1970), 'Mendición de algunos aspects del desarrollo socioeconómica de las entidades federativas de México, 1940–1960', *Demografía y Economica*, 292–322, Page 310.

1965 Gross Internal Product—Carrillo Arronte, R. (1972) 'Regiones geoeconómicas de Mexico', *Mercado de Valores*, **XXXII**, 234–43.

Peru 1961 National Income—Banco Central de Reserva del Perú, (1966) *Cuentas nacionales del Perú 1950–65* (Lima), Pages 35–37.

Philippines 1948, 1960, 1966. Estimated Gross Regional Product—Sicat, Gerardo P. (1968) *Regional economic growth in the Philippines* (Manila, Institute of Economic Development and Research, University of the Philippines: Discussion Paper No. 68–1), Pages 36, 42, 45.

Spain 1955–70 Per Capita Income—Banco de Bilbao, *La distribución provincial de la renta*, various years.

Tanzania 1963 Gross Domestic Product—Bureau of Resource Assessment and Land Use Planning (1968) Research Paper No. 1. *Regional economic atlas mainland Tanzania*, Pages 69–71.

1967 Gross Domestic Product—*Tanzania second five-year plan for economic social development 1 July, 1969–30 June 1974, Volume III Regional Perspectives* (Dar es Salaam, 1970)

Thailand 1960–69 Gross Domestic Product—Regional Accounts and Data Analysis Unit, National Accounts Division, NEDB (1970) *Gross domestic product—Northeast Thailand 1960–1969.* Pakkasem, P. (1972) *Thailand's Northeast economic development planning: a case study in regional planning* (Doctoral Thesis, University of Pittsburgh).

Venezuela 1969 Gross Regional Product—Travieso, F. (1975) *Ciudad, región y subdesarrollo* (Caracas: Fondo Editorial Común), p. 97.

Chapter 6

Growth-Centre Strategies in Less-Developed Countries

Jaya Appalraju and Michael Safier

The spatial organization of development and change in Third World countries is shaped by two major sets of forces. The first consists of those continuing influences and patterns of relationship that are derived from colonial or peripheral status, external dominance and internal underdevelopment. A second set arises from the post-war, post-colonial, political economy of individual Third World countries. The latter are expressed in deliberate public policies intended to reshape inherited organizations of space in line with prevailing government priorities for social and economic reconstruction. The concept of a 'growth pole'—however this may subsequently be defined—is becoming, in the form of growth-centre strategies of spatial or regional planning, one of the most widely promoted policy initiatives invoked by national governments aimed at reinforcing the impact of the second set of forces.[1] In this brief review our aim is to analyse the intellectual bases and practical potentials of growth-centre strategies as policy instruments in development planning; and in particular to examine the likely limitations of such strategies in many of their existing forms to bring about significant changes in the organization of national space. This entire area of discussion—of 'growth-centres', of 'spatial organization', of 'development and change'—is so alive with methodological controversies, poor conceptual understanding and empirical lacunae that in a short essay we can only contribute a small measure of clarification and open up leads to be followed in subsequent investigations.

Growth-centre strategies have long been practised by governments of less-developed countries without being dignified by the name as such. For example, the construction of the Owen Falls dam project of the then colonial administration of Uganda and the planned development of Jinja as an industrial and commercial centre to contrast with the administrative capital, Kampala, predate by several years the publication of Perroux's original articles on 'pôles de croissance' (Wilson, 1967; Larrimore, 1959; Perroux, 1950; 1955). It has been during the past decade or so, however, that growth-centre policies have become incorporated with growing explicitness into the national development plans of an increasing number of countries (Ford Foundation, 1972). It is of particular interest to note that countries where scarce capital, manpower and organizational resources have been committed to regional and spatial planning, involving one or more strategically conceived growth poles, points, centres or foci, vary widely in area and total population, in level of per capita income, in their degree of urbanization and industrialization, and in their political constitution.

In such countries, and in others where growth-centre policies are designated in name only, these policies are officially presented as efficient, effective and progressive responses to the need for organizing a new 'geography of development' that will reflect the economic, social and political aspirations of national governments. What is being suggested is the modification and reorganization of national or regional space, however hesitant and aparently confused the intention and however scanty the empirical diagnosis of prevailing conditions. The potential of growth-centre strategies to bring about such changes in the structure of spatial and regional relationships is clearly apparent in the basic formulations discussed in the next section. The actual achievement to date, in every country on which we have evidence to hand, however, is extremely limited. Existing national geographies of development and underdevelopment, the prevailing spatial organization of population, production and power, apparently remain subject to other more powerful influences.

The explanation for this gap between promise and delivery is found in a combination of three conditions, stated here in a steeply ascending order of importance. Firstly, the creation and still more the execution of growth-centre policies has been relatively recent in most countries. Significant changes in spatial structure are likely to take longer to mature than the period covered by one or two five-year plans. Nevertheless, if what is being attempted is more than small incremental adjustments to established economic and geographic patterns of activity, it should be possible to identify movements away from these patterns at a quite early stage in the production of an alternative economic and spatial formation. The signs are still few and far between.

Secondly, the general literature, which provides both the conceptual and theoretical bases for the elaboration of specific growth-centre strategies and the basic equipment for a critique of related policies, is at present of limited assistance in devising and testing national and regional programmes. Since their first formulation by Perroux in the middle fifties (Perroux 1950; 1955; Boudeville, 1966) notions of growth poles and growth centres have multiplied apace. Successive attempts to introduce more precise definitions and theoretical refinements (Darwent, 1969; Moseley, 1974) have left unresolved questions and often generated more confusion on every level of argument pertinent to development-planning strategies. We are still groping for even limited agreement on five crucial stages of policy making, viz:

1. The conception of what it is that constitutes a 'growth centre'.

2. The explanation of why it is that such a centre possesses particular qualities conducive to structural change in economy, society and space.
3. The specification of strategic objectives which can be furthered by the designation of any given centre or set of centres.
4. The policy instruments which can be used to implement centre development.
5. The empirical criteria for verifying and evaluating different forms of growth-centre development.

Dissatisfaction with our present state of knowledge and understanding in all these and other respects is evident in recent reviews which have attempted to bring some order and consistency to the existing literature (Darwent, 1969; Hermansen, 1972; Todd, 1973; Moseley, 1974). Further light on the range and content of growth-centre strategies being considered in less-developed countries, and on the conceptual and practical element involved has emerged from a number of international and national surveys, which have gathered together views and evidence so far available (Kuklinski, 1972; UNRISD, 1971-1975; EFTA, 1974; Misra, Sundaram and Rao, 1974). As a result of this recent writing the substance of both empirical and prescriptive foundations for public policy and planning has been considerably strengthened.

Thirdly, and most decisively, we must recognize the evolution—or revolution—in thinking and practice in development planning and in political economy. As has already been remarked, both in its basic notion and in its policy presentation the growth centre is a focus for the deliberate engineering of related structural changes in economy, society and space. The intrinsic nature and historical circumstances of most less-developed countries and prevailing conditions of poverty, inequality and underdevelopment offer a difficult environment in which to establish a centre or set of centres able to modify existing structural relationships in production, distribution and organization. It has become gradually evident that most spatial and regional planning policies which incorporate growth-centre strategies contain a contradiction in terms. In general, the designated centres have not been foci for structural transformation in the social formation of space, but rather have been encapsulated within, and contributary to, the pre-existing geography of development and change. Moreover, when viewed in terms of the realities rather than the rhetoric of governments in many less-developed countries, this contradiction is seen as consistent with a preferred continuity and extension of existing economic and social structures.

A better understanding of the effort required to bring about decisive changes in the geography of development, and of the political committment needed to bring about such changes as part of a radical reconstruction of national economy and society, has evolved in several Third World countries. It is precisely in these conditions that the successful execution of growth-centre strategies are most apposite and crucial.

The following pages attempt to link together four distinct but related tradi-

tions of enquiry and experience, as a means of placing growth-centre strategies within the context of the less-developed countries. The four components of our argument are:
1. The notion of 'polarized' development as a basic form of economic and spatial organization and transformation.
2. The historical and spatial analysis of development and underdevelopment.
3. A critical review of ideas and experience in national development planning.
4. An examination of 'types' of growth centres and their potential contribution to development and change.

In the next section we outline a general 'form' of the growth-pole idea incorporating a growth-centre translation, which we perceive in the relevant context as a 'tool' for implementing basic changes in economic and spatial structures. In the following section we trace the evolution of prevailing patterns of economic and spatial organization in less-developed countries, to whose radical transformation a growth-centre strategy is presumed to contribute. We then turn to an examination of development planning and its gradual progress towards more substantive definitions of structural change, and relate this progress to the adoption at various times of different types of growth-centre policies. Finally, we outline the cases of three highly distinctive countries and governments, Kenya, Tanzania and Iran, which have recently adopted comprehensive growth-centre strategies of potentially more lasting consequence.

THE TOOL: A GENERALIZED NOTION OF POLARIZED DEVELOPMENT

Having so far carefully avoided defining our terms, we will now argue for a single, very simplified, version of the multifarious growth-pole and growth-centre concepts. The logic of both the previous vagueness and the following extreme reduction is the same. In our view the main requirement of policy makers in less-developed countries is an easily understood notion of growth-centre development which embraces a variety of realistic spatial planning approaches but which has in common the concept of 'polarized' development. But what is the nature of this general operational idea, and how is it extracted from the intellectual discussions on the subject now available after twenty years debate?

In a recent article Moseley (1973) has graphically displayed the plethora of names and notions which surround the concepts of the growth pole and the growth centre. He demonstrates that authors writing in the last decade have used the terms poles and centres to encompass a very wide range of situations and places. Some have been concerned with the general nature of 'development', economic activity and demographic dynamics; some with the specifics of infrastructural investments and commercial services. Some have emphasized contributions to maximizing growth of medium or large cities; others with the establishment of small towns. Some writers have been interested in explanations of how existing concentrations of population and production have come about and others with prescribing an optimal pattern for future city systems.

This variety of ideas and terms has received contributions from many different intellectual perspectives, the historiography of which would make a revealing study in its own right. Our concern here however is to abstract certain elements common to most specifications of 'polarized' development and spatial organization. We have been considerably assisted in this by Moseley's (1974) monograph which attempts to bring some coherent order to the field.

We propose to specify the 'form' of polarized development as a single complex phenomenon involving six interrelated conditions, incorporating both growth-pole and growth-centre concepts. We observe however, as previously noted, a by now well-marked distinction between those conditions which apply to the activities involved in a pole and those that apply to the kind of spatial organization, or centre, which embodies the pole.

The conditions that formally characterize a growth centre for the purposes of policy making and strategic choice in less-developed countries appear to be as follows:

1. The presence of specific production, distribution and trade activities which have achieved, or are established at, some minimum scale of operation, and which have inherently, or externally introduced, technical or organizational qualities which give them a potential for rapid and sustained growth.

2. The evolution or establishment of direct and indirect linkages between activities included under (1) and other groups of activities which are, or can be, incorporated into a 'mix' or 'complex' of complementary and reciprocal exchanges.

3. The growth of activities included under (1) at a sufficient rate and on a sufficient scale to 'propel' the system of associated activities included under (2) in such a manner as to progressively 'transform' the structure of production and trade, involving the adoption of successive innovations and the cumulative addition of more specialized production capacities.

4. The evolved or induced organization of activities and processes included under (1) to (3) within a distint locational concentration having a degree of relative advantage and/or centrality in the national space and regional development.

5. The growth or initial establishment of locations included under (4) to achieve a minimum size sufficient for the realization of externalities and agglomeration effects additional to those involved in (1) to (3) above, and for the efficient emplacement and operation of supporting infrastructure and urban facilities.

6. The ability of a concentration of activity included under (4) at a scale of operation included under (5) to induce socio-economic development of surrounding areas and settlements, through trading, communication and other mechanisms of diffusion.

The first three conditions distil the central elements of a growth pole in economic development, the idea of an evolving or a planned activity sector which transforms an existing structure of production relationships and consumption patterns by incorporating within an economic system novel and more productive combinations of technology and organization. The subject has been a matter of discussion and controversy among theorists of economic growth and development since the seminal article by Rosentein-Roden on Eastern European development in 1943, but the *locus classicus* of the field lies in the writings of Perroux (1950, 1955, 1964).

His initial conception of a growth pole was as a means to explore and highlight the complex processes whereby clusters of economic activities and industries appear, grow and stagnate in an economic system over time. A pole is identified as a point in abstract economic space to which centripetal forces are attracted and from which (in time) centrifugal forces emanate throughout the field of influence of the set of activities constituting the pole. The field of a particular pole will exist at any period of time within a series of such fields surrounding other poles in various stages of evolution and interrelationship. It is intriguing in the context of contemporary controversies over 'dependence' and 'multinational' firms in the international economy to note that Perroux's idea of 'abstract economic space' was set in contrast to the more conventional idea of 'geographic (national) space'. The latter he considered to be of very limited relevance for the analysis of modern economic systems (Perroux, 1950).

The growth-pole conception of a group of firms or an industry which are primary generators of new productive forces and interrelationships emphasizes a particular kind of theoretical orientation in the explanation of economic growth and development. The impetus that shifts the larger part of an economic system away from an existing (equilibrium) state, and the propulsive mechanism that then underpins its dynamic development, is a series of disjointed thrusts involving the emergence of a series of poles of sufficient magnitude and potential to affect a wide range of contemporary production and consumption relationships. The emergence of a new pole will depend on the establishment of one or more of five 'enterpreneurial innovations':

1. The introduction of a new 'good' or of a new quality of existing goods.
2. The introduction of a new technology of production or commercialization.
3. The opening up of a new market area or reservoir of consumption.
4. The exploitation of a new source of supply of raw materials or manufacturers.
5. The achievement of a new monopolistic position.

In this perspective it is the ability to adopt innovations and capitalize on their potentials by concentrated effort and investment that serves as the major motive force in development. Individual growth poles over time will become cumulatively more powerful engines of change, the dominating nodes from which developmental impulses spread. A series of 'leading sectors' founded on major innovations will exploit all manner of scale economies and externalities, and gradually reshape the productive relations of older and lesser poles around its own marketing and demand requirements. The greater the innovation and rate of growth achieved, the more significant will be the impact on the total economic landscape.

These basic notions are closely related to a group of ideas expressed in contemporary writings on growth and development in the middle fifties. In particular, Scitovsky's article on external economies (1954), Rostow's concept of 'take-off to self-sustained growth' (1963), Myrdal's model of 'cumulative causation' (1957) and Hirschman's (1958) thesis of 'unbalanced growth', all contributed to and reinforced the notion of 'polarized' development in the macro-economic sphere. By pointing to this wider intellectual inheritance we are choosing to emphasize the 'structural' dimensions of growth-pole analysis and strategic planning and the notion of departure from a dominant pattern of productive and institutional relationships within an economic and social system. It is an emphasis to which we will return.

The second group of conditions, (4) to (6) are distilled from parallel discussions among theorists of spatial organization and regional development. The most direct equivalent that has been drawn between a growth pole as defined in (1) to (3) and a locational or spatial arrangement is the one suggested by Boudeville (1966) and extended by Lasuén (1969). A large urban centre becomes established on the basis of intrinsic or acquired locational advantages and is able to attract specialized and potentially dynamic activities which provide a self-sustaining momentum of growth and change. Through the changing locational interdependencies of a hierarchically ordered 'system of cities' the new growth centre 'organizes' a new spatial structure of production and interregional trade and diffuses a range of innovations throughout its surrounding area. In turn, this process of organization and diffusion will transform the existing patterns of population and production concentration, cumulatively reinforcing or transforming the inherited national space-economy.

Around this basic notion can be gathered a spectrum of related perspectives which both deepen and broaden analytical and operational conceptions of growth-centre development. At one end of the spectrum are detailed and precise analyses of technically and organizationally powerful 'activity complexes' or 'territorial production complexes', the immediate localized interdependencies of specific concentrations of rapidly growing leading sectors (Isard, 1959, 1960). Further along are studies of the way in which new growth centres are formed within the environment of existing systems of cities and their impact on the overall development of such systems (Berry, 1972). Further still are the more generalized approaches to the nature of polarized development in space and its relationship to transformations in the structure of economic and social organization which lead on to the identification of different types, levels and stages of centre evolution, and their converse, the emergence of peripheral areas and regions (Friedmann, 1966; Hermansen, 1972).

Such brief comments and references can do no more than suggest the substantive intellectual antecedents of a generalized notion of polarized development such as we have proposed here. The considerable and continuing controversy in the reviews to which we have referred cannot be justly treated in this presentation. Nevertheless, the concept of the growth pole embodied in a growth centre does not possess in its generalized form sufficient identity for the elaboration of a variety of policy initiatives intended to modify, and in some cases replace, existing geographies of development and change. The concentration of effort and investment in one or a few strategic locations, in order to achieve a substantive and sustained breakthrough to a new pattern of areal change and influence, has been embodied in a range of spatial and regional policies each of which is designed to react against contemporary conditions and conventional modes of operation. To appreciate the impulses and limitations of such deliberate measures we must turn to an examination of the historical circumstances in which they were reared.

THE TASK: SPATIAL ORGANIZATION AND STRUCTURAL TRANSFORMATION

When growth-pole theory was being developed in the late fifties, there was scarcely any coherent account of 'the geography of development and change' within which to appraise the potential contributions and limitations of any proposed growth-centre strategies. Since 1960, however, two lines of work have emerged. First, the more systematic investigation of relationships and structures which contribute to the ordering of people and activities over space with particular reference to 'developing' countries (Keeble, 1967; Johnson, 1970); and second, the work on processes of development and underdevelopment as expressed in sequences of spatial organization (Soja, 1968; Riddell, 1971; Gould, 1970; Gilbert, 1974). These developments have provided a dual set of ideas and evidence on which to build our present outline.

Over any given area, the distribution of population and economic activities and the physical movements and interactions which connect them, are complicated, irregular and apparently random. By initially breaking down the total complexity of spatial structure into several constituent components it is possible to trace the underlying regularities. The geographic analysis of development and change has attempted to account for alterations in the configuration of people, production and power which occur over time, by reference to elements such as the location of economic activities, the spacing and size of settlements, the direction and intensity of movements, the networks of transportation and communications, the spread of innovations and the process of decision-making within corporate organizations. Each and all of these elements can be related to prevailing economic and social formations and their associated technologies, and then related together to characterize a matching 'spatial organization'.

In brief, movements of people and commodities follow observable tendencies, the volume and direction of movement being related to magnitudes of activity, i.e. progressively greater between areas having large populations and high levels of production. This relationship is extended and shaped by both the 'friction of distance' and the 'attraction of complementarity': migrants move from areas where opportunities (real and perceived) are scarce to those where they are abundant, while goods and services are exchanged between centres of production and areas of demand. Other things being equal, the greater the distance, the higher the costs of moving and the less perfect the information available, the less will be the flows involved. Movements reflect the existing order of population and production over an area, and accomplish the transfer of people and goods responding to that order. Movements are conditioned by the network of routes, which, in turn are determined by the spatial configuration of population and production. The degree of 'connectivity' of a place, measured by the ease with which other places can be reached via a direct routing, is closely aligned with the 'centrality' of a place, i.e. its relative distance from the 'mass' of population and activity located in a given area. In turn, the web of routes further support the arrangements of people and activities by providing the means of minimizing the friction of distance in the transfer of both.

In addition, movements on networks are closely related to the location of centres or 'nodes' of particularly intense settlement and activity, which are themselves related together in some form of 'hierarchy'. The organization and exchange of goods and services between groups of people over an area is carried on by way of a regular series of functionally specific centres which are 'spaced out' according to population and production densities and levels of productivity, within which there appears a graded structure of centres in ascending order of size and specialization. Taken together these nodal centres form a system through which the distribution of goods and services is mediated, from the most diverse production locations to major points of consumption and exchange. This regular pattern of centres is complicated by the evolution of specialized production locations which are at the same time major consumption centres, and thereby incorporate a large part of the support given by the system of centres to minimizing transport and other costs by means of collapsing or 'agglomerating' population and production within a localized segment of the total area. The overall densities of population and levels of production of different categories, related to available movement opportunities, transport networks, central places and agglomerations, are further consolidated through specific elements of comparative advantage and accessibility. There exist over any given area a complicated and intertwined series of 'distance differentials' which affect the relative density of occupation and activity by their actual and potential contributions to satisfying local and area-wide demands for goods and factors of production. Thus the 'surface' of population and activity and 'levels' of life and welfare are a reflection of underlying configurations and forces—a skin stretched over the framework of networks and nodes which gather to themselves the bulk of movements and interactions which supply the needs of economic and social exchange. They are the product of a composite of conditions, historic, physical, technical and organizational, which work together to make some areas or regions comparatively more attractive to new growth than others.

We can use this 'spatial accounting' scheme to illustrate the widespread evolution of systems of spatial organization connected with historical stages of development and underdevelopment. Despite great variations in most aggregate characteristics such as overall size of area, population densities and such like, most 'pre-colonial' societies relied on using local materials and resources to satisfy local needs, had a territorial-based socio-economic structure, and were bound in allegiance to local chiefs and kings. The intensity of economic exchange and social interaction were limited by distance and technologies even within the confines of a single ethnic grouping, external connections remaining peripheral, made up of sporadic trading, depredation or conquest. The multifaceted relationships and close interdependence of communities closely identified with a particular locality bred sharp discontinuities in contacts and exchanges over large distances. Certain important exceptions to this generalization occurred where more specialized patterns of internal exchange and external trade were long established, such as during the West African empires where there existed a higher level of political centralization, or with the Inca domains in Andean America, or where migration into a new ecological setting gave rise to new styles of life and livelihood as repeatedly happened in southwest Asia. Otherwise, however, the frequency and range of movement for goods and persons, the elaboration of a settlement structure, and the levels of intensity and specialization in localized clusters of activity, all remained slight.

The initial impact and subsequent evolution of a colonial economy and society involved the elaboration of a totally new type of spatial organization. Within any given area there were created extreme specializations of activity with high intensities of operation, with similar types of production scattered over widely separated regions and in different parts of the world. New social institutions and norms of behaviour were spread throughout the colonized territories, with specialized roles and novel institutions created for each local area. A centralized political authority held sway over comparatively vast areas, with a newly created set of internal administrative divisions and external boundaries defining the exclusive colonial space. International exchange became the basis of commercial production, with the various specializations in resource-cum-labour exploitation becoming focused on a new set of central places which drew on distant sources to satisfy both local and export demands. A structure of economic, social and political relationships was reproduced which bound the local populations into a web of externally organized contracts between seller and agent or client and official, on terms which were dictated by the possession of capital, technology and force. Increasing levels of intraregional specialization and interregional connectivity were complemented by interregional migration flows, and structured by networks of transport and communications which hinged on centres of coordination and control arranged in a strict hierarchical order. At the apex of this system were created one or a few superordinate foci mediating the new patterns of exchange and terms of trade within and between the widespread peripheral hinterlands. The deification of a new 'modern' order was accomplished by the diffusion of technical innovations and education, and the parallel destruction of all local competition (Safier, 1968; Gould 1960, 1970).

In the newly independent nations of Africa and Asia, and in their older established Latin American counterparts, the pattern of export orientation, primate-city domination, and transport networks focused on the main ports, is a common inheritance. The limited scale, and simple character, of many post-colonial national economies, and the complementary social formations which have evolved to work them, have continued the concentration of investment and other resources which reinforce prevailing patterns of spatial organization. This condition has stimulated successive governments in a wide range of developing countries to attempt the redefinition of their 'geography of development' so as to complement their efforts to achieve new, and less 'dependent', development goals. Such redefinitions have given a spatial dimension to ideas of modifying or replacing inherited and distorted patterns of resource use, and of maximizing human and physical resource use for the benefit of the mass of the local population. Unfortunately, the most appropriate form of spatial organization for Third World countries is far from apparent. Despite, or because of this problem, growth-centre strategies have been widely canvassed as excellent instruments for governments to reorientate internal production and organizational interlinkages towards the sustained and more equal development of local resources. Conceptions of how such strategies are to be used in this task have been gradually evolving over the past twenty years roughly in line with changing views on the priorities of development policy. It is from the latter that we can gauge the actual and potential effectiveness of growth-centre strategies in combatting the powerful combination of inertia and dominance in spatial structures that currently prevails.

THE RESPONSE: GENERATIONS OF DEVELOPMENT PLANS AND THE SUCCESSION OF GROWTH-CENTRE PROJECTS

The evolution of ideas on the realities of development and the limitations of development policies has been a gradual and piecemeal process. Clearly, however, it has had a decisive effect on the direction and content of growth-centre strategies, firstly as modifications to, and more recently as reactions against, current patterns of spatial distribution. In both cases, it has been the interaction of new ideas and new experience—and principally the unsatisfactory nature of both–that has been the driving force towards more novel and comprehensive formulations.

Sufficient attempts at formulating and implementing national development plans have now been accumulated in the Third World for some tentative generalizations to be made. A recent study of some twenty less-developed Commonwealth countries identified no fewer than 150 development plans and equivalent documents published between 1947 and 1974 (Koenigsberger and Safier, 1973). These plans can be classified into four distinct but overlapping 'generations', each involving important shifts in objectives and priorities, and reflecting a constant expansion in the role of government in the structuring of the national space-economy.

In a large number of former colonial dependencies, the first proper development plans were preceded by one or more of a generation of reports on 'Postwar Reconstruction' or 'Development and Welfare'. These resulted from directives to government ministries to submit long-term (up to ten year) capital budgets and consisted of lists of public-sector investments for which capital loans or grants from an external source were sought. The first attempt to move outside prevailing colonial practice was the generation of 'Public Sector Investment Programmes'. These contained for the first time overall development objectives and explicit policy priorities, conventionally for a five-year period, and specified production targets, investment allocations and intersectoral projects. They were still restricted in the main, however, to those sectors operated or controlled by the government. Their perspective on development stressed the facilitative role of public sector programmes in lending support to commercial enterprises so that infrastructural improvements, agricultural extension and education provision were repeatedly emphasized. With the exceptions of India, Pakistan and Sri Lanka (then Ceylon) such programmes and perspectives were predominant throughout Africa and Asia until the end of the fifties and indeed remained quite common until the late sixties. Typical examples of this species are Malaya's First Five-year Capital Expenditure Plan (1956–1960), and Trinidad's Five-year Economic Programme (1950–1955).

A next generation of plans, which began to appear on a significant scale in the early sixties, was associated with the attainment of political independence and followed the model established on the Indian subcontinent ten years before. This generation of plans was the product of an urgent desire to force the pace of economic growth, and to alter the basis of access to economic power in favour of the new governments, and those they represented. These plans, therefore, embodied a decisive rise in the expected level of national investment, in the range of guidance and intervention in both public and private sectors of the economy, and in the number of output targets and organizational adjustments affecting resources and incomes. Such 'National Economic Development Plans' are still the classic form of plan undertaken in Third World countries: prominent examples being India's First Five-year Plan (1951–1956) and the First Plan of Pakistan (1955–1959).

By the late 1960s or early 1970s, however, the perspective on development had been further clarified. In a small but increasing number of countries there was a growing awareness of, and concern for, the negative structural, social and political implications of the by then conventional 'growth-orientated' development strategy. This was especially evident in those countries where such strategies had failed to overcome the most pressing problems of poverty, unemployment and inequality, or to substantially increase the productive capacity of the domestic economy vis-à-vis the dominance of large-scale international capital, technology and market organization. This latest generation of 'Comprehensive National Development Plans', therefore, attempts to specify targets and identify instruments, not only to stimulate the growth of major productive sectors, but also to deal with sensitive subjects such as manpower planning, agrarian organization, unemployment, income distribution and the external ownership of local assets and resources. Recent examples of this type of plan are Tanzania's Second Five-year Plan for Economic and Social Development (1969-1974), Malaysia's Second Plan (1970-1975), and Trinidad's Third Five-year Plan (1969-1974).

The evolution of growth-centre strategies as instruments for the reshaping of development geography has followed a similarly partial and piecemeal course. It is only quite recently and in a few countries that a more comprehensive conception of such strategies, embedded in a radical departure from previous unidirectional assumptions and consequent policy actions, has come to the fore. We cover three such cases in the next section. For the most part, the successive emphases of growth-centre policies have taken the form of single, albeit major, projects, embedded in a wider framework of economic and spatial organization. On this framework they have been able to exert no decisive influence, in large part because their potential has never been realized to the extent defined in our combinations of elements (1) or (6) on page 147 We can identify six different categories of growth-centre 'projects', which overlap in conception and in their relation to the three main 'generations' of development planning referred to above. Each category of growth-centre embodies within its conception the general notion of polarized development but each introduces a new collection of leading sectors and propulsive forces expected to achieve the particular development objectives valued at the time.

The first class of growth-centre strategies concentrated on the provision of *infrastructure*. Within the earliest perspective on development which stressed the relief of basic impediments as the key to the growth of directly productive enterprise, the initial growth-centre concept was couched in terms of public sector, pump-priming operations to provide a critical minimum level of power, water, transportation and other amenities. An early example of this class is the one quoted at the beginning of the paper—the establishment of a hydroelectric power source at Jinja in Uganda. The leading sector of this centre was to be the power facility and the closely associated copper-processing operations. A complex of related activities was to include a range of medium-scale, electric-powered, manufacturing plants producing mainly consumer goods, and the centre's rationale, and propulsive force, was seen as the introduction into a primary-producing economy of a diversified secondary sector which would stimulate both agricultural processing and manufacturing to satisfy local consumption. The focus of locational concentration was established immediately adjacent to the power source, in the geographical centre of the most developed region of the country. Agglomeration and threshold economies were to be secured by attracting all 'national' manufacturing investment over a ten or twenty year period, and the parallel expansion of the town to 200,000 inhabitants or more. Finally, extensive development throughout the surrounding area was to be induced by stimulating agriculture to produce a food surplus for urban consumption, backed up by complementary material and organizational inputs from the urban industrial economy that would progressively transform rural and interregional interlinkages.

It is but a short intellectual step from here to the second and third classes of growth centre, the overlapping categories of *industrial* and *regional* centres. However, in policy terms, the step is wider since the intentions are usually more radical and the aim to reshape the structures of national or regional economy and society. The former class is represented by the development of Tema in Ghana, a port and industrial centre adjacent to the capital, Accra; the second by the case of Ciudad Guayana in a remote 'resource-frontier' region of Venezuela; and a mixed case by Durgapur, Asansol and the related cluster of centres in the Bengal-Bihar heavy industrial region of northeast India. Industrial centres have been intended to generally upgrade the embryonic structure of national manufacturing, while regional centres have been established to transform the existing pattern of regional activities, usually on the basis of exploitable resources for industrial production. In the heyday of national development policies aimed at accelerated industrialization, both industrial and regional centres had much the same basic logic and 'content'. The leading sector in each case was to be one or more large-scale, heavy or intermediate manufacturing, operations established by public sector investments—iron and steel, aluminium, petrochemicals or heavy engineering—with an associated 'industrial complex' of component and complementary plants. The whole project was to produce a propulsive force sufficient to transform intersectoral linkages via the incorporation of more complex technologies into the domestic economy. The focus of concentration was established close to sources of basic raw materials and/or at transport foci (such as sea or river ports) with good links to national or regional markets. Agglomeration and threshold economies were to be secured by the sheer magnitude of the industrial and urban activities involved; urban populations were planned to reach a level of between one-quarter and half-a-million people within ten or fifteen years. In the case of regional centres, development was to be induced in the surrounding and undeveloped hinterland through the attraction of local migrants to jobs in the new complex, and through access to specialist services and support facilities newly available within the region.

These first three classes of growth-centre project have a common emphasis on the direct use of large-scale investment resources to generate structural changes through accelerated economic growth and the resultant technological and organizational spin-offs. Experience has taught, however, that both the macro-economic and the macro-spatial 'environments' of most developing countries contain severe limits to the effectiveness of polarized development projects of this kind. Two basic kinds of constraint have been the restrictive size and 'absorptive capacity' of small markets for industrial outputs, coupled with the capital-intensive, low employment-generating character of much of the industry involved. In addition, these centres have had a severely limited impact in transmitting developmental impulses through their surrounding areas because the linkages involved in the development of the centre itself have been largely with 'external' suppliers and markets, and because the derived demand for labour and for agricultural produce has stimulated migration and supplies from 'outside' the relevant region in which the centre is located. The overall rate of growth of output, and still more of employment and population, has been slowed by these circumstances, and the generation and diffusion of locally-inspired innovations, able to 'take hold' of local client groups and local resources, has been minimal. The total costs involved in establishing each such centre moreover have been so considerable in terms of total investment and other resources, that their relative failure to make a rapid and extensive impact on prevailing structures of production and spatial organization has made it increasingly unlikely that similar centres will follow.

Partly in reaction to the experience of growth-orientated development strategies and the types of growth centre associated with them, a fourth and fifth class of centre has become more widespread—the *rural* and the *service* centres. The fourth type is represented by the development of 'agrovilles' in Pakistan, the fifth by the establishment of 'new towns' in the Kelantun and Trengganu provinces of West Malaysia, and a mixture of the two by the designated central places for 'intensive (rural) development zones' in Zambia, or the 'rural growth centres' of local government (block) areas in selected Indian states. Rural and service centres are intended to transform the structure of production and settlement on a local but more widely spread basis than their predecessors. The leading sector is seen as one strategically important component in the relevant range of input-output improvements or complementary facilities needed for modern commercial agriculture and/or efficient non-agricultural rural production. The complex of related activities embrace other locally needed components in the same range of activities—farm input provision, marketing facilities, processing plants, component manufacture, repair services, local education, health schemes and the like. The propulsive force of this range of newly accessible elements in rural improvement is meant to derive from the simultaneous or phased provision of facilities, incentives and opportunities which can transform the means, the practice the organization and the direction of rural economies and societies. The locational concentration of rural and service centres is based on the spatial allocation of a limited number of core units of productive or service capacity over an area, the addition of the whole range of 'central-place' functions and localized rural industries, and the attraction and settlement of population up to the size of a small town in which a minimum level of urban infrastructure is economically viable. In this case the socio-economic development of the surrounding areas is induced by the upgrading of agricultural practices and non-agricultural activities which is expected to take place through the interrelationships of the centre and its surrounding settlements. It is the very multiplicity and intricacy of the necessary 'leverage' on the local socio-economic structure that has made the rural and service centres so difficult to sustain and limited in effect. Without a more powerful stimulus from external market opportunities, experience has repeatedly been of a slow and localized series of changes in levels of productivity and adoption of new ideas. While a 'concentrated dispersal' of public sector services and facilities may have been achieved, the complementary mobilization of local resources, skills and initiative has proved hard to induce.

The sixth and latest class of growth centre has emerged in the form of the *metropolitan* growth centre; the 'twin-city' plan for the Bombay metropolitan area in India is an example. The logic and content of such a centre are concerned with the planning and management of city regions and the modification of an existing monocentric pattern of population and activity concentration. The leading sector of such centres is provided by the decanting of one or more central city functions—office development, large-scale industry, commercial and transport services—and the gathering around this of all manner of related support and derivative activities which are expected to capture a substantial proportion of metropolitan economic dynamism unhampered by the existing constraints on the pre-existing central city. The new metropolitan centre is located at a strategically advantageous point within the city-region, has ample land for expansion and good access to interregional communications, and is planned to expand its productive and residential development to city size and beyond in a decade or two. As against nearly all the previous classes of centre the economic and spatial environment for metropolitan centre growth would seem to be uniquely positive; but though experience with this type of centre is still limited it would seem that the organizational, managerial and financial complexities of such an endeavour, plus the powerful 'dynamic conservatism' of established central city complexes, inhibits the successful promotion of an alternative focus well before it can reach a level of self-sustaining momentum sufficient to bring about the transformation in land-use and residential configurations it was designed to achieve.

The record of achievement of growth-centre strategies, conceived as simple 'projects' and executed within the wider environment of economic and spatial organization that continues to characterize contemporary underdeveloped countries, has been somewhat limited. Fortunately, a further generation of experiments with growth centres appears to be emerging which is based on the idea of a multiplication of centres and the complete reorganization of the national space-economy. We turn to a consideration of this new departure in the following section.

THE FUTURE PROSPECT: NATIONAL URBAN POLICY AND SYSTEMS OF CENTRES—KENYA, TANZANIA AND IRAN

By the end of the 1960s it was becoming clearer that the conditions of poverty, inequality and underdevelopment still frustrating the planned progress of many Third World countries were deeply embedded in their external relations and internal economic and social formations. As such, they remained untouched by accelerated rates of economic growth and by the execution of any single-sector or project priorities however dramatic in themselves. Similarly, the promotion of growth-centre projects (of whatever 'class') were seen as incapable of achieving the full potential of polarized development without at the same time being part of a more comprehensive revision of national distributions of population and production. In recent years, a few countries have designed new policies for the establishment and promotion of a national 'system' of growth centres. These policies have aimed at the consolidation or the reorganization of the total national space. In almost every case, the use of the growth-centre form has been suggested for the further development of a middle level in the urban hierarchy, the so-called 'secondary' cities. These cities are the regional capitals and interregional foci of production and trade that together make up an emerging urban system that in time may replace the present unidirectional spatial structure with its 'primate' agglomeration and related metropolitan dependence. They are, on the one hand, the pivotal centres necessary for a national system of cities with the capacity both to mobilize local resources for local needs within individual regions and to articulate interregional exchanges of specialized products and services. On the other hand, they provide the sites for the organization of an effectively decentralized form of development administration and management which may achieve the more efficient allocation and use of national resources.

Countries having different varieties of political economy and social formation, and with different policy objectives have chosen to emphasize different mixtures of 'contents' for a national system of growth centres. We here examine three distinct types of system: that based on the service-centre hierarchy being evolved in Kenya; that based on regional centres being promoted in Tanzania; and that based on the industrial centre being employed in Iran. In all three cases, however, the following strategic components are found in common.

1. The designation of a 'system' of centres relating to the country as a whole.
2. The direct relation of that system to the overall national development strategy.
3. The recognition of a commitment to the designated system of centres lasting beyond a single plan period (of normally five years).
4. The forecast and/or specification of future (population) sizes for the individual centres and the allocation of investment funds for their further development.
5. The concept of the system of centres having a multisectoral or multifunctional concentration of productive activies and supporting facilities and services.
6. The placing of the system of centres in the context of national urban policies relating both to primate centres and to other levels of spatial organization.
7. The direct association of growth centres with primary levels of decentralized decision-making and resource allocation in carrying out the national plan.

These characteristics of the new systems of growth centres add up to a more

Table 6.1 Past, present and projected populations of systems of growth centres

Country	Urban Populations ('000s)		
1. *Kenya*	(1962)	(1969)	(1980)
Nairobi	347	509	1,100
Mombasa	180	247	447
Sub-total	527	756	1,547
Kisumu	24	33	124
Kakamega	4	6	17
Eldoret	20	18	42
Nakuru	38	47	79
Thika	14	18	50
Nyeri	8	10	26
Embu	5	4	14
Sub-total	113	136	352
2. *Tanzania*	(1959)	(1967)	(1974)
Dar es Salaam	128	272	462
Sub-total	128	272	462
Tanga	38	60	102
Moshi	14	27	47
Arusha	10	32	78
Mwanza	20	35	58
Morogoro	15	25	42
Dodoma	13	23	39
Tabora	15	21	31
Mbeya	8	12	23
Mtwara	10	20	28
Sub-total	143	255	448
3. *Iran*	(1956)	(1966)	(1977)
Tehran	1,510	2,700	5,110
Sub-total	1,510	2,700	5,110
Tabriz	290	405	580
Arāk	59	73	123
Esfahān	255	423	720
Ahvaz	120	207	370
Ābādān	226	273	336
Shīrāz	170	270	447
Mashhad	242	409	710
Sub-total	1,362	2,060	3,286

Source: All figures derived from respective Development Plans and projections by the author on the basis of official census and other accounts.

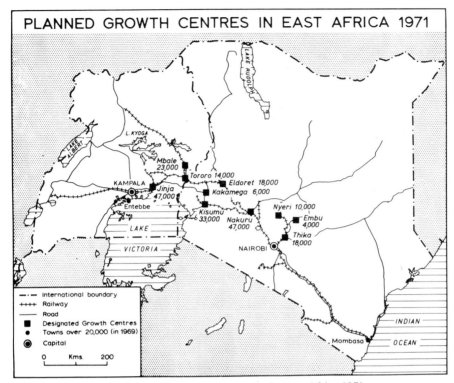

Figure 6.1 Planned growth centres in East Africa 1971

powerful and concerted thrust towards the reshaping of spatial organization on a national scale than has been observable hitherto. With reference to the relevant development plans, and with the aid of Table 6.1 and Figures 4.5, 6.1 and 6.2 we can indicate the direction of this thrust in three contemporary cases.

Kenya's Second Development Plan (1970–1974) (Kenya, 1969) sets out the overall perspective for the physical planning of national resources and settlement, and the role and function of the designated 'major growth centres'. A brief review of the current distribution of urban and regional development in the country leads on to the outline of a strategy for the decentralized development of urban service functions as a basis for the better distribution of secondary activities and the simultaneous stimulation of agriculture and rural improvement. The overall national strategy will incorporate the continued expansion of the dual foci of the existing urban system—the capital, Nairobi and the port of Mombasa—and the designation of four levels of a settlement hierarchy corresponding to local district development in a classic 'central place' formulation. In between these levels there appear a small number of existing major towns with some degree of manufacturing and commercial specialization; these are to become the future secondary cities and form the system of centres which will service major segments of the national space. They have the qualifications of already being important industrial and administrative centres, of holding strategic economic locations, and of possessing substantial infrastructural and service facilities. Kisumu and Kakamega will command the west of the country and the densely settled Lake Victoria basin; Eldoret and Nakuru the rift-valley uplands; and Thika, Nyeri and Embu the heavily populated highlands immediately northeast of Nairobi itself.

Kenya's Third Development Plan (1974–1978) (Kenya, 1974) amplifies the concept of the 'planned provision of urban services and infrastructure throughout the country' by making special reference to the spatial and physical planning required for the long-term evolution of each major urban centre. It notes, however, the difficulties being experienced, on the one hand, in executing the required development against the continuing *ad hoc* nature of project location decisions, and on the other, in providing the needed urban facilities and services in line with productive sector expansion and population growth. In this context, the significance of the growth centres or 'principal towns' is further elaborated in relation to the complementary development of rural areas and the absorption of rural migrants, the more even spread of accessible higher-level services, and the articulation of a national transportation network which will stimulate interregional trade.

Tanzania's Second Plan for Economic and Social Development (1969–1974) (Tanzania, 1970) sets out a different perspective on the aims of overall development and the spatial planning of rural development and related secondary and tertiary activities. In contrast to Kenya's acceptance and strengthening of an existing pattern of urban development, the Second Plan emphasizes the objective of reducing urban primacy and regional inequalities by a dual thrust towards integrated rural development and the redistribution of economic growth over a number of secondary centres and away from the capital, Dar es Salaam. It is recognized that its natural and acquired advantages will secure for Dar es Salaam a continuing position of prominence in the national system of cities and in our view will lead to further concentration of manufacturing and administrative activities in the primate city and to the parallel deprivation of the mass of rural people. The response, in line with the policy priority established in the now famous 'Arusha Declaration' for rural reconstruction and the promotion of economic and social equality, is the elaboration of a carefully researched programme of rural and regional investment for the whole country. An important ingredient in this programme is the promotion of secondary cities to act as centres for regional administration and service provision, and as the foci for a decentralized industrial location strategy aimed at expanding job opportunities outside agriculture in all areas of the country. A number of the designated 'nine towns' are already established commercial and manufacturing centres with hinterlands of advanced commercial agriculture—the northern port of Tanga, the twin towns around the Kilimanjaro Massif, Arusha and Moshi, and the Lake Victoria port of Mwanza. The others, how-

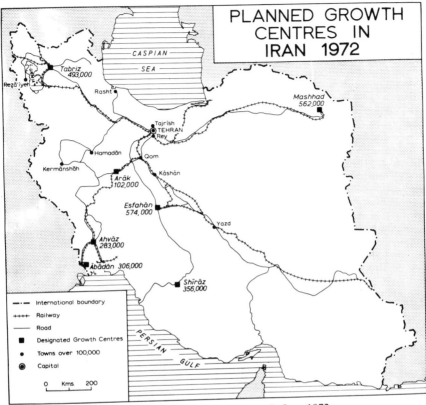

Figure 6.2 Planned growth centres in Iran 1972

ever, are district centres in the centre and south of the country in areas of extreme poverty and underexploited resources, where the promotion of urban growth will provide a much needed stimulus and, if successful, will reorientate the existing system of cities towards a more equitable national coverage.

Determination to radically reconstruct the space-economy of the country in line with the socialist transformation being prepared in economic and social relationships has led to a further and dramatic departure at the end of the Second Plan period in Tanzania. The functions of the administrative and political capital are to be removed from Dar es Salaam over a ten-year period beginning in 1975, and sited at Dodoma, one of the most central of the regional growth centres and one which is located in the poorest district of the country. As in the case of Arusha, where the establishment of administrative headquarters for the East African Economic Community and its related organizations doubled the target population inside a decade, the establishment of national government authorities in Dodoma is expected to double its 1974 population by 1980. Along with the construction of the Tanzania-Zambia rail and road connections, this will further strengthen the new system of regional centres as countermagnets to the old established concentrations of development on the coast and in the north.

Iran's Fifth Development Plan (1974–1978) (Iran, 1974) records a different perspective to either of the East African cases. Here the emphasis is on the creation of a national system of cities and towns able to organize the spatial location of activities in those sectors which are being stimulated by the huge increases in oil revenues. In particular, the plan incorporates multiple objectives into the elaboration of spatial policy, ranging from the provision of adequate infrastructure to contain the upsurge of productive and welfare sector operations, through the careful husbanding of natural resources and continued modernization of rural life, to the securing of a better distribution of benefits and facilities to all parts of the country. A central plank in the resulting programme is the specification of a number of existing major urban areas as growth centres within which to concentrate major national investments, particularly industry but also education, tourism and so on. These centres are expected to attract large-scale state and private enterprise away from Tehran; to stimulate the mobilization of the natural resources of their particular regions; and to constitute a system of secondary cities drawing closer together the hitherto very distinct regional economies and population groups stretching from the border with Turkey and Russia in the northwest to the margins of Pakistan and Afghanistan in the southeast. Each of the designated centres has been allocated massive investment funds for a coordinated programme of heavy industrialization, round which are expected to cluster numbers of lesser units, boosting individual city populations by two-thirds or more in a decade. The urban industrial specializations are a role call of the main elements in the establishment of a modern industrial capacity: engineering and transport machinery at Tabriz, engineering and machine tools at Arāk, iron and steel and textiles at Esfahān, petrochemicals at Ābādān, metal manufacturing and chemicals at Ahvaz, light engineering at Shīrāz and vehicle assembly and textiles at Mashhad. This is growth-pole and growth-centre strategy exercised in a systematic manner on the grand scale.

It is as yet too early to judge whether the three systems of growth centres reviewed here—and others which will appear shortly—will soon show signs of achieving any or all of the objectives of national urban policy and related spatial reorganization which they are intended to serve. There is a strong case, however, for attempting to monitor and analyse the ongoing efforts now being made, in the reasonable expectation that they will provide a new and interesting development in the theory and practice of growth centres. If the kind of careful empirical work done by Moseley (1974) and others, on the character and use of growth centres in developed areas, can be extended to such cases as presented here, then we may soon expect answers to our questions about the capacity of polarized development to provide national governments in less-developed countries with a tool for restructuring their economic and social arrangements and their associated patterns of spatial organization.

NOTES

1. In the discussion that follows we reserve the term 'growth pole' to refer to an aspatial, economic concept concerned with development dynamics (after Perroux, 1950; 1955) and the term 'growth centre' to cover a locational and spatial concept related to the above.

REFERENCES

Agarwala, Amar N., and Singh, Sampat, P. (Ed.) (1963) *The economics of underdevelopment* (London: Oxford University Press).
Arrighi, G., and Saul, John S. (1973) *Essays on the political economy of Africa* (New York: Monthly Review Press).
Berry, Brian J. L. (1972) 'Hierarchial diffusion: The basis of development filtering and spread in a system of growth centres', in Hansen (1972), 108–138.
Blaug, Mark (1964), 'A case of the Emperor's new clothes: Perroux's theories of economic domination', *Kyklos*, **XVII**, 551–564.
Boudeville, J. R. (1966) *Problems of regional economic planning* (Edinburgh: Edinburgh University Press).
Brookfield, Harold (1975) *Interdependent development* (London: Methuen).
Cameron, G., and Wingo, Lowden (Ed.) (1973) *Cities, regions and public policy* (Edinburgh: Oliver and Boyd).
Darwent, D. F. (1969), 'Growth poles and growth centres in regional planning—A review', *Environment and Planning*, **1**, 5–32.
European Free Trade Association, Secretariat (1974) *National settlement strategies: A framework for regional development* (Geneva: EFTA).
Ford Foundation (1972) *International urbanization survey reports:* Volumes on Chile; Colombia; Kenya; Nigeria; Peru; Turkey; Venezuela; Zambia (New York: Ford Foundation).
Friedmann, John (1966) *Regional development policy: A case study of Venezuela* (Cambridge, Mass.: M. I. T. Press).
Gilbert, Alan (1974) *Latin American development: A geographical perspective* (Harmondsworth: Penguin).
Gilbert, Alan (1975) 'A note on the incidence of development in the vicinity of a growth centre', *Regional Studies*, **8**.
Gould, P. R. (1960) *The development of the transportation pattern in Ghana* (Evanston, Ill.: Northwestern University Studies in Geography, No. 5).
Gould, P. R. (1970) 'Tanzania 1920–63: The spatial impress of the modernisation process', *World Politics*, **22**, 149–170.
Hägerstrand, T., and Kuklinski, A. (Ed.) (1971) *Information systems for regional development—A seminar* (Lund: Lund Studies in Geography, Series B, No. 37, Gleerup).
Hansen, N. M. (Ed.) (1972) *Growth centres in regional economic development* (New York: Free Press).
Hansen, N. M. (1967) 'Development pole theory in a regional context', *Kyklos*, **XX**, 709–727.
Hermansen, Tormod (1972) 'Development poles and development centres in national and regional development', in Kuklinski (Ed.) (1972), 160–203.
Hirschman, Albert O. (1958) *The strategy of economic development* (New Haven, Conn.: Yale University Press).
Iran (1974) *Fifth Development Plan: Summary* (English) (Teheran: Government Printing Office).
Isard, W. (1959) *Location and Space Economy* (Cambridge, Mass: M.I.T. Press).
Isard, W. *et al* (1960) *Methods of Regional Analysis* (Cambridge, Mass: M.I.T. Press).

Johnson, E. A. J. (1970) *The Organization of space in developing countries* (Cambridge, Massachusetts: M. I. T. Press).
Keeble, D. E. (1967) 'Models of economic development', *Models in Geography*, Chorley, R. J. and Haggett, P. (Eds.) (1967), 243–302.
Kenya, Republic of (1974) *Development Plan, 1974–1978: Part I* (Nairobi: Government Printing Office).
Kenya, Republic of (1969) *Development Plan, 1970–1974* (Nairobi: Government Printing Office).
Koenigsberger, Otto H., and Safier, Michael S., with Fallon, Kathleen (1973) *Urban growth and planning in the developing countries of the Commonwealth: A review of experience from the past 25 years*—Paper presented at the Interregional Seminar on New Towns, London, July 1973 (London: United Nations).
Kuklinski, A. (ed.) (1972) *Growth poles and growth centres in regional planning* (The Hague: Mouton).
Larrimore, A. E. (1959) *The alien town* (Chicago: University of Chicago, Department of Geography Research Paper 55, Chicago University Press).
Lasuén, J. R. (1969) 'On growth poles', *Urban Studies*, **6**, 137–161.
Lasuén, J. R. (1971) 'Multi-regional economic development: An open system approach', in Hägerstrand & Kuklinski (Eds.) (1971), 169–211.
Mabogunje, Akin (1971) *Growth poles and growth centres in the regional development of Nigeria* (Geneva: UNRISD Report, N. 71.3, United Nations).
Misra, R. P. (1974) *Spatial framework for multi-level perspective planning in Iran, with particular reference to Fors Ostan* (Teheran: Centre for Research & Training in Regional Planning/United Nations Development Programme).
Misra, R. P., Sundaram, K. V. and Rao, V. S. L. P. (1974) *Regional development planning in India: a new strategy* (Delhi: Vikas Publishing House).
Moseley, M. J. (1973), 'Growth centres—a shibboleth?', *Area*, **5**, 143–50.
Moseley, M. J. (1974) *Growth centres in spatial planning* (Oxford: Pergamon Press).
Myrdal, Gunnar (1957) *Economic theory and underdeveloped regions* (London: Duckworth).
Pakistan, Planning Commission (1971) *Fourth 5 Year Plan, 1970–75* (Islamabad: Government Printing Office).
Perroux, F. (1950) 'Economic space, theory and applications', *Quarterly Journal of Economics*, **LXIV**, 90–97.
Perroux, F. (1955) 'Note sur la notion des pôles de croissance', *Economie Appliquée*, **1** and **2**, 307–320.
Perroux, F. (1964) *L'economie du XXe siècle* (Paris: Presse Universitaire de France).
Pioro, Zygmunt (1972) 'Growth poles and growth centres: theory as applied to settlement development in Tanzania', in Kuklinski (Ed.) (1972).
Riddell, J. B. (1971) *The spatial dynamics of modernization in Sierra Leone* (Evanston, Ill.: Northwestern University Press).
Rosenstein-Rodan, P. N. (1943) 'Problems of industrialization of Eastern and South-Eastern Europe', in Agarwala and Singh (1963), 245–255.
Rostow, William W. (1963) 'The take-off into self-sustained growth', in Agarwala and Singh (1963), 154–186.
Rweyamamu, Justinian (1971) *Underdevelopment and industrialization in Tanzania: A study of perverse Capitalist industrial development* (London: Oxford University Press).
Safier, Michael S. (1968) 'Urban growth and planning in Africa: 1 Economics, 2 Geography', in *Papers of the University of East Africa Social Science Conference* (Kampala: MISR).
Safier, Michael S. (Ed.) (1970) *The role of urban and regional planning in national development for East Africa—Papers and Proceedings of a Seminar* (Kampala: Milton Obote Foundation).

Saul, John S. (1973) 'African Socialism in one country: Tanzania', in Arrighi and Saul (1973), 237–335.
Saul, John S., *et al* (1972) *Towards Socialist planning in Tanzania* (Dar es Salaam: Tanzanian Publishing House, Tanzanian Studies No. 1).
Schumpeter, J. A. (1934) *The theory of economic development* (Cambridge, Mass.: Harvard University Press).
Scitovsky, T. (1954) 'Two concepts of external economies', *Journal of Political Economy*, **62**, 143–151.
Soja, E. (1968) *The geography of modernization in Kenya* (Syracuse, N. Y.: Syracuse University Press).
Tanzania, United Republic (1970) *2nd Five Year Plan for Economic and Social Development, 1st July, 1969—30th June, 1974—Volume III: Regional perspectives* (Dar es Salaam: Government Printing Office).
Thomas, M. D. (1972) 'Growth-pole theory: an examination of some of its basic concepts', in Hansen (1972), 50–81.
Todd, Daniel (1973) *The development-pole concept and its application to regional analysis: An appraisal of the state of the art* (London: London School of Economics, Graduate School of Geography Discussion Papers, No. 47).
Tornqvist, G., and Pred, Allan R. (1973) *Systems of cities and information flows* (Lund: Lund Studies in Geography, Series B, Human Geography, No. 38, Gleerup).
Trinidad & Tobago (1970) *Third 5-Year Plan, 1969–73* (Port-of-Spain: Government Printery).
Uganda, Republic of (1972) *Third 5-Year Development Plan, 1971/2–1975/6* (Entebbe: Government Printing Office).
UNRISD (1971–75) *Reports on Regional Development* (The Hague and Paris: Mouton).
Wilson, Gail (1967) *Owen Falls* (Kampala: East African Publishing House).
Wingo, Lowden (1973) 'Issues in a national urban development strategy for the U.S.', *Urban Studies*, **9**, 3–28.

Chapter 7

Urban Concentration and Dispersal: Urban Policies in Latin America

J. A. S. Ternent*

INTRODUCTION

Economic development and industrialization have been held to be closely associated with urbanization. Such an optimistic view has reflected the experience of the developed nations where a lengthy period of rising average per capita incomes was accompanied by a major exodus from the rural areas. The settlement pattern produced by this process was marked by the emergence of contiguous built-up areas. The rapid increase in agricultural productivity which marked this development, together with the concomitant release of productive factors, especially labour, from the rural areas, took place at a time during which industry was still labour intensive. Furthermore, as population growth was relatively slow (at least by present-day standards), displaced rural labour found employment in the towns which grew as a result of local diversification and sustained development (United Nations, 1970, Kuznets, 1966, pages 114–115). The transition from a predominantly agrarian, rural society to an urban, industrial one was therefore possible, if not particularly easy.

That this may well have been a special case is to be inferred from the more extensive experience with urbanization in parts of West Africa and the Middle and Far East where old and well-established cities have coexisted with little appreciable economic development over considerable periods of time. In fact, these cities appeared to one observer as 'a planner's nightmare and a mass of seemingly insurmountable problems' (Hahn-Been, 1968, page 9). During the present century this kind of view has naturally suggested dispersal as the appropriate development policy to be followed.

In addition emphasis on the economic efficiency of investment in urban infrastructure has been due to the past squandering of building resources in Eastern Europe's largest cities. Today, therefore, exports from both the Soviet Union and other Eastern European countries stress the need to optimize city size in order to limit the growth of these large metropolitan areas. Preference

---

*The views expressed here do not necessarily reflect those of the General Secretariat of the Organization of American States.

is given to the development of selected intermediate and small urban centres as well as that of urban systems. Besides the Soviet Union, the systematic stress on limiting the growth of large cities is evident in a number of important declarations for cities such as Budapest, Prague and Warsaw (Stván, 1973, pages 10–11, 57–58).

The rapid growth of Latin American cities, particularly the large ones, since the Second World War, together with the high cost of providing public services, has similarly led to the prescription that cities other than the metropolitan areas be stimulated. While it may be overstating the case to assert that a consensus on the need for dispersal has been reached, over sixty regional programmes within individual countries in Latin America have recently been documented; in addition fourteen programmes involving more than one country are under way (Stöhr, 1972). It may be stated, therefore, that there is considerable interest in programmes focusing on areas other than the large cities. Although natural resource development has been an important factor underlying these programmes, urban development has also been affected in so far as they have modified settlement patterns, particularly by strengthening the smaller cities. Underlying these strategies seems to be a feeling that the large cities do not contribute to development as they should. This would not necessarily conflict with the 'cities as engines of growth' thesis deduced from Western European and North American experience, for during their periods of rapid growth cities were still relatively small. However, two strands of thought now stand out: the first relates urbanization to economic development whereas the second is concerned with the relationship between development and primate versus non-primate settlement patterns.

The lack of success encountered by past dispersal policies and the apparently inevitable growth of the large cities, together with the finding by anthropologists and architects that most urban migrants adapted well to city life, combined to produce during the 1960s a re-examination of the role of the large city with a view to isolating its positive contribution to national economic and social development. A number of conferences examined the leading role of cities in economic development (due largely to the fact that the more dynamic segments of industry were located within them) and the more obvious unfavourable factors associated with excessive growth (such as high and growing unemployment rates, congestion and pollution, as well as a variety of social problems and their physical expression in low quality urban settlements). The objectives of the United Nations Interregional Seminar on Development Policies and Planning in Relation to Urbanization, held in Pittsburg in 1966, were to find means of exploiting the positive aspects of urbanization while minimizing the undesirable effects of overcentralization of population and industry. In other words, urban policies and programmes were to be integrated into overall national plans for economic and social development (United Nations, 1968, page 4). Similarly the focus of the Cornell Conference on 'The Role of the City in the Modernization of Latin America' was on the positive aspects of the city in the development process rather than on the 'usual urban problems'. The aim was to identify those components of urbanization that can be accelerated without unduly aggravating the normal problems associated with rapid urban growth (Beyer, 1967). Implicit in this view is the feeling that large city growth has produced a set of difficult problems rather than making a positive contribution to the welfare of its inhabitants. While the unfavourable factors may well be held to be symptomatic of the fact that cities are permanently in transition which, together with the conception of development as a unilinear evolution, suggests that such difficulties will inevitably be overcome (Culbertson, 1971), the very rapid growth of already large Latin American cities suggests that they will rapidly encounter, if indeed they have not already done so, a series of barriers to orderly future growth which will be very costly to overcome. One alternative is, of course, to countenance cities made up of ever larger lower income groups living under unacceptable conditions. Emphasis on the linkages between national development, change, and cities as primary agents of social, economic and political development has caused interest to progress 'from preoccupation with the problems of the growing city to an intensive examination of its positive role as an instrument of modernization and national development' (Hahn-Been, 1968). Yet another, and surely more positive, alternative consequently lies in acceptance of the fact that both settlement patterns and living conditions within particular cities are amenable to public policy. It is surely not simple coincidence that the Ecuadorean provinces of Pichincha (Quito) and Guayas (Guayaquil) had the highest rates of immigration in Ecuador between 1950 and 1962 while receiving on average over 50 percent of public sector investment between 1955 and 1964 (Ternent, 1974). Similarly heavy emphasis on the construction industry by the Colombian authorities during 1973, especially in Bogotá (the capital city), has reportedly cut open unemployment there in half (from 11.3 percent of the labour force in 1972 to 6.3 percent by the end of 1973).

Recent experience in Brazil, particularly that of the São Paulo region strongly suggests the importance of factors other than public policy in channelling migratory flows, and consequently on the formulation of national policies for urban growth. One such indicator is the coincidence between an accentuated process of economic development and the emergence of a relatively large number of 'new towns' in the region (that is, towns which surpass the 20,000 inhabitant mark). It is surely not accidental that the São Paulo state accounted for a full one-fourth of the increase in the number of cities for Brazil as a whole over the period 1950–1970 (IDB, 1972; IBGE, 1971).

In response to events such as the above, interest has centered on what has come to be known as national urbanization policies, recognizing the role of the city explicitly in national development. Thus, unlike prior conferences which dealt with the internal structure of the Latin American city, the Jahuel (Chile) seminar emphasized the national significance of urban growth, nationwide urban systems, and the socio-economic and political integration of urban areas with surrounding peripheral regions (Miller and Gakenkeimer, 1969). Even more comprehensive was the objective of the United Nations Interregional Seminar on the Financing of Housing and Urban Development (Denmark, 25 May–10 June 1970) according to which human settlements are to be viewed, and their growth and development planned, 'within a national and regional framework which would be integrated with the economic, social and political planning process' (United Nations, 1972). Similarly the role of urbanization in development, impact on the environment, size of city, decentralization, urban–rural balance, and the dualism which marked cities in the developing world, were major themes at the Rehovot Conference on Urbanization and Development in Developing Countries (*Interim Report*).

Perhaps the most comprehensive statement of this view was made in conjunction with the United Nations Conference on the Human Environment (Stockholm, 1972). 'The confrontation and juxtaposition of the performances of the man-made and the natural components generate conflicts which affect the (human) settlement's complex eco-system and its inhabitants. An essential condition for the survival and future development of humanity is the reduction and ultimate elimination of these conflicts. In order to eliminate them, it is necessary to establish a balance among the economic and social aims of development and the physical forms which these take. Such a balance can best be achieved by the coordination of government policies and activities through comprehensive planning at national and regional levels' (United Nations, 1971, page 7). While there can be no doubt but that economic development and urbanization have been positively related (United Nations, 1968, pages 21–35), the introduction of environmental considerations as suggested above leads to the conclusion that the relationship is not linear. The use of measures of urbanization which purport to more adequately reflect a country's settlement pattern suggests that it is less the relationship between development and urbanization than the interplay between development and the pattern of urbanization which acquires significance from a policy point of view (Arriaga, 1970; Ternent 1974). It is with this general issue that this chapter is concerned.

The next section will examine the literature concerned with patterns of urbanization and the formulation of public urban policy. In particular it will describe the statistical analyses which suggest that there is an eventual and natural tendency towards deconcentration or dispersal of urban growth as development proceeds. Such a reversal need not, however, coincide with the most desirable moment from a policy point of view. Consequently in the following section the policy implications of allowing very large cities to grow are examined. It should be clarified from the outset, however, that no universally optimum city size is being postulated. In the final section some attempt to sketch out an urban policy is made. This would appear essential given the large and rapidly growing contingents of lower income groups who concentrate in the large cities; the presence of significant regional income and cost differentials within the context of resource constrained economies in Latin America, the need to revitalize agriculture and provide alternatives for migrants from the rural areas; and the need to improve large-city environments. The need now, therefore, seems to be for general instruments of analysis which, when applied to particular cases, yield appropriate strategies for them. The final section is consequently intended as a modest contribution to the implementation of public policies seeking a balance between the needs of large and small cities, within a context of national economic development.

DEVELOPMENT AND THE PATTERN OF URBANIZATION

The first proposition to be examined is that national patterns of urbanization bear an identifiable relationship with levels of economic development, that is, as a country develops, changes in the spatial distribution of its population may be predicted with reasonable accuracy. The discussion has focused on development in a country over time as a transition from a disequilibrium situation characterized by a primate city distribution (where the population in the main city is more than twice that of the next largest city) at a early stage of development, to the postulated equilibrium of the rank-size distribution found in many developed countries. That this is the resulting equilibrium has been argued on both statistical and probabilistic grounds (respectively entropy maximization, and the end result of a random arrangement of a given number of people in a given number of cities).[1] Empirical analysis of this proposition has yielded the two-part conclusion that, while primacy does indeed appear to be related to several characteristics of small less-developed countries (such as low per capita income, high export dependence, a colonial history, an agricultural economy, and high rates of population growth), half the wealthier countries in one sample exhibited primacy while the other half did not (Linksy, 1965). Subsequent studies either found only partial support for this conclusion (Mehta, 1964) or rejected it outright (Berry, 1964 and Mills, 1971). However, not only has the validity of deriving correlation coefficients from heterogeneous sets of countries been questioned but the need for caution in generalizing from one culture to another emphasized, especially on the basis of limited 'Western' experience (Sovani, 1964). Furthermore, the analysis in terms of two polar cases has encouraged the denial of any relationship between primacy and development because some underdeveloped countries have been found to be non-primate in their city size distributions while some developed countries were primate (El-Shakhs, 1972, page 14).

More recently, however, the testable hypothesis has been advanced that decentralized spatial systems may be characteristic of both very underdeveloped and wealthy countries if several development stages are recognized, e.g. traditional, transitional and developed (El-Shakhs, 1972). This view is based upon a sequential theory of core-periphery relations according to which polarization and dependence increase with modernization up to a point at which intensified conflict or tension between the core and the periphery results in a reversal of the trend towards concentration. Clearly the factors at work in the first and last stages would be different and, where the latter is concerned, there does not seem to have been much progress beyond an expressed desire to know more about the reasons why primacy characterizes some developed

Table 7.1 Ranks according to different indices of development

	Development El-Shakhs	Index UNRISD	Rank El-Shakhs*	Rank UNRISD**
United Kingdom	496	104	1	2
Belgium	484	96	3	8
Germany (Fed. Rep.)	493	94	2	11
United States	478	111	4	1
France	474	88	5	12
Switzerland	458	96	6	8
Canada	456	103	7	3
Netherlands	448	96	8	8
Sweden	446	103	9	3
Australia	443	98	10	6
Italy	438	71	11	20
New Zealand	430	103	12	3
Japan	429	74	13	16
Norway	427	98	14	6
Poland	411	73	15	18
Finland	400	85	16	16
Hungary	391	75	17	15
Argentina	398	73	18	18
Spain	387	58	19	23
Mexico	380	44	20	30
Venezuela	371	63	21	21
Yugoslavia	370	51	22	26
Chile	363	61	23	22
Portugal	357	52	24	24
Uruguay	355	74	25	16
Israel	343	81	26	14
Brazil	335	38	27	31
Egypt (U.A.R.)	325	34	30	40
Greece	323	52	29	24
Turkey	320	27	30	40
Morocco	292	26	31	41
Peru	286	37	32	32
Costa Rica	281	50	33	27
El Salvador	266	32	34	34
Philippines	260	24	35	43
Panama	257	48	36	28
China	254	46	37	29
Guatemala	250	21	38	45
Ecuador	243	31	39	36
South Korea	241	25	40	42
Dominican Republic	227	30	41	37
Nicaragua	222	32	42	35
Honduras	215	23	43	44
Thailand	201	10	44	46
Paraguay	187	29	45	38
Libya	139	29	46	38

Source: *El-Shakhs, 1972, p. 14.
**UNRISD, 1970, p. 151; Davis, 1969; IDB, 1972.

Table 7.2 Primacy and development

	Primacy Index		UNRISD Development Index
	El-Shakhs	Davis	
United Kingdom	.709	1.51	104
Germany (Fed. Rep.)	.703	0.96	94
Belgium	.689	0.75	96
United States	.697	0.88	111
France	.690	3.57	88
Switzerland	.634	0.71	96
Canada	.732	0.68	103
Netherlands	.664	0.51	96
Sweden	.741	1.14	103
Australia	.678	0.71	98
Italy	.665	0.56	71
New Zealand	.673		103
Japan	.783	1.62	74
Norway	.701		98
Poland	.749	0.62	73
Finland	.694		85
Hungary	.736	4.60	75
Argentina	.741	4.15	73
Spain	.731	0.90	58
Mexico	.739	1.74	44
Venezuela	.735	1.70	63
Yugoslavia	.698	0.64	51
Chile	.745	3.50	61
Portugal	.834		52
Uruguay	.808		74
Israel	.735	0.86	81
Brazil	.764	0.72	38
Egypt (U.A.R.)	.773	1.91	34
Greece	.903		52
Turkey	.739	1.18	27
Morocco	.808	1.40	26
Peru	.882	5.32	37
Costa Rica	.860	3.67	50
El Salvador	.907	2.50	32
Philippines	.802	4.27	24
Panama	.888	3.04	48
China	.600	0.63	46
Guatemala	.932	6.21	21
Ecuador	.741	1.19	31
South Korea	.751	1.09	25
Dominican Republic	.830	3.47	30
Nicaragua	.861	2.83	32
Honduras	.792	1.64	23
Thailand	.890		10
Paraguay	.894	5.68	29
Libya	.654		29

Source: See Table 7.1

countries but not others. Here, however, interest centres on the less-developed countries.

The statistical procedure followed in testing the above proposition consisted in developing an indicator of primacy and relating it to an index of development for a sample of 75 countries (all information being for the year 1955).[2] The initial conclusion was that a negative correlation exists between the degree of primacy and the level of development (Tables 7.1 and 7.2). Inspection of the scatter diagram suggested, however, that the observations fall into two separate groups. While the more developed countries appeared to confirm the negative correlation mentioned above, a positive correlation was found for the less developed (El-Shakhs, 1972, page 22). Formal analysis confirmed this suggestion, for the coefficients of less-developed and developed countries were positive and negative respectively and higher than the coefficient obtained for the entire sample. Primacy, therefore, may be expected to rise until a midpoint in development is reached; subsequently it will decline.

While statistical difficulties were encountered with both the development and primacy indicators,[3] the use of alternative indices[4] confirmed the negative slope found above although the correlation was weak.[5] However, inspection of the new scatter showed no tendency for the countries to split into two groups. Interestingly, the developed countries for which the new indicators were available clustered quite well around the best fit line. Below the average level of development, however, there was no apparent relationship at all for all the countries included. For Latin America, on the other hand, there may well be an upward drift if the two most primate countries, Guatemala and Paraguay according to this indicator (no data were available on Uruguay), are discounted.

Recent historical and statistical research enables further light to be cast on the above results (McGreevey, 1971). Information on the city-size distribution in preindustrial or 'traditional' societies is now available for several Latin American countries.[6] A distribution statistically similar to the lognormal or deconcentrated pattern has been identified as an initial state for these countries during the 18th century. Mexico was the only exception, primacy having developed there as early as 1750. Systems of cities may not have been present, however, and these early distributions in the Latin American colonial states have been attributed largely to high transportation and administration costs. In other words, Latin America at the time was composed of a series of isolated regions and approximated the large country case in which primacy is invariably low. In this historical analysis it is also hypothesized that the growing 'openness' of the Latin American economies (as measured by per capita exports) in great part accounted for the continuously increasing divergence from the lognormal distribution, i.e. the emergence of primacy, during the 19th and 20th centuries. The recent growth of primacy can only be termed explosive.

Use of a recently proposed operational measure of concentration[7] also supports the reversal hypothesis outlined above although this may be due to the fact that the concentration measure was chosen precisely on the basis of the highest correlation with a set of development indicators (see Table 7.3).

Table 7.3 Comparative primacy ranking in Latin America

	Primacy according to El-Shakhs		Primacy according to Arriaga	
	Index	Rank	Index	Rank
Argentina	.741	14	117	2
Mexico	.739	15	23	11
Venezuela	.735	16	44	5
Chile	.745	12	77	3
Uruguay	.808	9	215	1
Brazil	.764	11	9	15
Peru	.882	5	32	7
Costa Rica	.860	7	42	6
El Salvador	.907	2	19	12
Panama	.888	4	75	4
Guatemala	.932	1	18	13
Ecuador	.741	13	18	13
Dominican Republic	.830	8	24	10
Nicaragua	.861	6	32	7
Honduras	.792	10	28	16
Paraguay	894	3	28	9

Sources: Table 7.1 and Arriaga (1970), pp. 216–217.

However, within the range of development exhibited by Latin America, a positive though diminishing relationship was found for Latin American countries; the index of primacy rose seven-fold while the index of development showed only a three-fold increase.

The positive correlation found between primacy and development for the less-developed countries should really come as no surprise, for several observers have suggested that a primate pattern emerges when development processes are left to themselves. There is much less unanimity with regard to the natural reversal of concentration. Once some critical level of development and concentration is reached, continued growth is felt to be compatible with widely different city-size distributions (von Boventer, 1971), although previous analysis of the behaviour of regional income disparities does provide indirect support for the reversal hypothesis. It has been suggested that regional disparities in income rise with development up to some midpoint, after which they decline (Williamson, 1965). This implies that since urban concentration is positively related to higher income levels, at least in the early stages of development, urban concentration is a necessary condition for economic development. Unfortunately, the positive relationship between urban concentration and development is increasingly felt to be unacceptable. The centre-periphery argument holds that large cities favour the accumulation of wealth during the early stages of development, even though living standards of large population segments may decline. That concentration should be typical of the 'take-off' is said to be due to the advantage lying with the developed centres, which exhibit the necessary overhead facilities, external economies, political power, spatial preferences of decision-makers, in-migration of the more vigorous and educated elements from the underdeveloped regions, flows of funds from large rural landowners to the financial markets in the cities, etc. Deconcentration may subsequently take place through the operation of natural mechanisms. These include improved access of secondary centres to innovation, wage increases relative to small cities, a diminishing return to investment, the spread of literacy and improved bureaucratic practices, the opening of transportation routes, and universal education and standardization (Alonso, 1968). The basic strategy is the pursuit of rapid income growth through exploitation of the most productive investment opportunities until income levels have reached such a point that a society can afford to worry about the well-being of its members.[8] Not only is continous and linear growth implicit in this view, but there is no consideration of whether in fact such natural growth diffusion mechanisms will work in desired ways. Thus, improved road access under the São Paulo action plan is said to have fostered the city's growth rather than spurring deconcentration (CINVA, 1970). Where the wage mechanism is concerned, rapid population growth is said to keep wages low in the metropolitan area (that is, lower than if there were a slower rate of population growth), inhibiting the transfer of activities out of the large into the smaller cities (Berry, 1971). A further mechanism is held to be a diminishing rate of return to investment in the large city although this has consistently been offset by growing external economies and innovation (Johnson, 1970). There is also a lack of information concerning the unfavourable effects of urban growth on the periphery and the varied advantages associated with large urban centres. If investment is in fact biased toward the larger centres, and contrary to some assumptions, wages are higher there than elsewhere, capital-intensive techniques will be favoured and there will be little incentive to move to smaller towns as relative wages rise. A further reason why the growth-diffusion mechanism may not function in the less-developed countries through the labour market is that unions impose a spatially invariant wage (Thompson, 1972). Thus wage differentials are kept small whereas productivity differences may be relatively large. The concept of a culturally acceptable minimum wage also militates against wages accurately reflecting productivity differentials. Internal dependence may also inhibit the growth-diffusion mechanism, and one objective of regional planning may well be to rationalize, though not eliminate, this relationship so as to serve the interests of the system and not just those of the dominant region (Boisier, 1972). The same issue may be expressed in terms of the allocation of productive resources between a metropolitan area and its hinterland in order either to maximize the welfare of the city ('colonialism') or the joint welfare of city and hinterland (Wingo, 1972).

    In sum, it would appear that there are a series of factors which systematically bias private location towards the large city and which, simultaneously, inhibit any degree of dispersal. While these will be explored further below, they include a lack of coordination in location decisions and imperfect foresight besides those mentioned above. The presence of external effects and, possibly, the favouring of metropolition areas by 'the whole complex of public finance and governmental policy' would contribute to making the primate settlement pattern even more unfavourable, in the sense of reducing national income from what it would have been under a more decentralized settlement pattern (Neutze, 1965).

Finally, to postulate continuous and inevitable economic growth is to ignore the time dimension in economic development; for even if a natural reversal in concentration of a country's population is to be expected, its timing need not be convenient from the point of view of a nation's development if its objectives include anything other than growth maximization. As most development plans incorporate multiple objectives, however, the reversal may be desired sooner, or even conceivably later, than would occur in the normal course of events. Such change in timing could be desired on the basis of objectives which are every bit as valid as that of achieving the most rapid rate of growth in per capita incomes.

The analysis of primacy developed above concentrated on relative magnitudes. However, absolute size is also of concern due both to an interest in the direction of causation and to give operational content to the natural reversal hypothesis. To the extent that primacy depends upon the various factors outlined above, the pattern of urbanization is seen as stemming from characteristics of the development process, in so far as physical factors are not the root cause. Causality may be reversed just as easily, making development (and urban economic growth) a function of the pattern of urbanization and the size of city. Thus, initial advantage and the overcoming of 'minimum' thresholds enables the benefits of economies of scale, agglomeration and urbanization economies, and indivisibilities to be achieved.

These factors are subsequently reinforced by a cumulative circular causation which may produce accelerated growth in a few cities, at least up to a point after which diseconomies of urban size slow urban expansion (Hirsch, 1973; Richardson, 1971). If a country's size distribution does tend to become less primate as development proceeds, and if development is itself a function of urbanization (a tenable hypothesis as the leading growth sectors tend to concentrate in the cities), then some at least of the nation's cities may be expected to advance through a series of stages paralleling to some extent the growth of the country within which they are located. Depending upon the mix of factors identified above, some urban centres will progress through the following stages: small city, export-specialization city, nodal centre, regional capital, and national capital. Each one of these stages will, be characterized of course by a smaller number of centres than the previous one. While such a movement is to some extent associated with population size, it perhaps depends more on differences in industry mix, export share of economic activity, demographic characteristics, and geographical location (Hirsch, 1973, page 284). Urbanization, therefore, depends as much on changes in the rank of cities (some ascending and others descending although this latter possibility would appear to be more associated with either developed countries or those in which population growth is proceeding slowly) as on an increase in the proportion of a country's inhabitants living in urban areas. For present purposes it may be hypothesized that early 'stages' of development are positively associated not only with urbanization, but with increasing city size and primacy, for the benefits of urbanization are most clearly felt with large cities. As shown by average city size and its associated measure of concentration, urbanization and development are positively related (as would be expected from a perusal of the literature) although a fairly wide range of concentration is evident (Ternent, 1975, Table A–2). This is probably due to the effect of size of country, for the larger the country the greater the chances of its having developed regional capitals. Of the six most developed countries in Latin America, the largest (Brazil and Mexico) show the lowest degree of concentration.

On balance, then the theoretical and empirical material summarized above suggests that primary rises with development, at least up to a point following which the city-size distribution tends towards the rank-size. The strongest statement to the contrary is that there is no relationship between the type of city distribution and either relative economic development or the degree of urbanization of countries, although urbanization and economic development are highly associated.[9] The view that the city-size distribution proceeds from simple or primate to complex or ranksize on the basis of 'the increasing complexity of economic and political life' can only hold therefore after the reversal (Berry, 1964, page 147). Furthermore, it has been pointed out that complexity is in many cases related to development and growth in the economy (El-Shakhs, 1972, page 13). Therefore, if in fact a simple and universal empirical relationship between these variables does exist, the weight of the evidence presented above is that primacy (as well as city size) is accentuated by development, although both voluntary and involuntary or natural factors may be involved in reversing this trend as tensions increase and become evident to policy makers. Data for Latin America suggest that policies designed to promote decentralization would also favour lower income groups (at least on average, for income levels and income inequalities in the small cities are lower than those observed in metropolitan areas). The pursuit of an improved regional distribution of income is, of course, one of the basic reasons underlying observed efforts in Latin America to achieve more balanced settlement patterns. The three-stage hypothesis is, therefore, very fruitful for it is surely not coincidental that deconcentration policies involving a slowing in the growth of the population in the largest centres coupled with an acceleration of intermediate city growth, are coming to the fore in many Latin American countries. Indeed virtually every nation that deals with this issue in its policy does so in the belief that its principal city is too big, regardless of its size (Alonso, 1971). While the social and political factors documented elsewhere may account for the emergence of such policies, it is clear that primacy is increasingly held to be disfunctional and 'gigantism' in the largest cities a characteristic to be feared (Berry, 1971).[10]

It would seem clear that as it stands the analysis is of little practical value. In order for it to become operational it is necessary to go beyond the reversal hypothesis and suggest both the conditions under which such a reversal will take place, as well as the size of city at which tensions begin to build up to a point at which some degree of deconcentration becomes politically viable. As this set of issues is closely related to the discussion of optimal city sizes and policies designed to bring about an improved use of space within a country, it will be discussed subsequently. No optimum city size is necessarily implied by the findings on a reversal in concentration. If indeed the former exists, it is an absolute magnitude whereas the entire argument concerning the relationship between urbanization and development is couched in relative terms, i.e. proportions of populations resident in cities of different classes are of importance.

Finally, a word of caution: it is so difficult to extract any very firm results from world-wide data that no particularly valid generalizations may even by present, at least at the level of aggregation used up to the present time. Alternatively the meagre results obtained may simply lend further weight to Sovani's (1964) admonition concerning the dangers inherent in cross-cultural comparisons. The Latin American data, on the other hand, do seem to support certain regularities when the available information is used either as a cross-section or in time-series form. To generalize from this to other regions may, therefore be quite risky.

SIZE OF CITY AND PUBLIC POLICY

In much the same way as the analysis of the relationship between urbanization and economic development suggests that, within a given national context, there is a desirable upper limit to concentration and city size, it should be possible from a review of the discussion of optimum city size to deduce some guidelines with regard to a national policy of urban growth. Such a policy may establish the priority cities in which to invest at a particular moment in time. In this fashion it may be possible to arrive at an urban strategy in which not only would specific elements of the urban environment be variables, but the cities themselves would also be subject to analysis from the point of view of the convenience of concentrating action within them. Thus, in the broadest interpretations of costs and benefits, it may be possible in the first instance to determine a strategy for public investment, as well as for other governmental activities, which would enable the location of population and economic activity to contribute more effectively to the achievement of national economic and social development goals. The purpose of this section, therefore, is less to argue that an optimal city size in fact exists than to glean policy conclusions from the literature on whether to consciously implement the 'natural' reversal in concentration observed previously.

Empirical research has mainly concentrated on the definition of proxies for benefits, costs, and city size or the degree of urbanization. While it may well be true that a city's population is not an unambiguous measure, depending as it does on geographic and political boundaries, advances are being made in two directions: on the one hand are the more precise definitions made possible by aerial photography which not only enable the contiguous urban area to be identified but to be kept up to date, especially with remote sensing now a practical reality; on the other are new concepts which enable other dimensions of urbanization to be brought to bear (Arriaga, 1970; Ternent, 1974). It is nonetheless true that absolute urban size as measured by the number of inhabitants in the established city limits, while useful for some purpose, remains a poor proxy for many other aspects, especially as it may imply widely dissimilar scales and mixes of public services due to different income levels in the cities (Richardson, 1972). However, the improved measures of urbanization and city size should enable some of these objections to be overcome, especially if they are modified to take income into account (Richardson 1969).

Some measure of income (real or monetary, total or per capita) has usually been taken to represent benefits in empirical studies. The general finding is that money incomes are higher in the larger cities (Hoch, 1972) and that these are characterized by higher real incomes because they are more productive (Wingo, 1972). Furthermore internationally it is asserted that there is no evidence that income increases diminish even for cities as large as New York and Tokyo (Mera, 1973). A typical example in Latin America is Mexico, where as Table 7.4 shows income and city size are positively related (Revista Mexicana de Ingeniería y Arquitectura, 1973).

This is by far the most damaging empirical finding where optimum city size is concerned. The burden of the observation is that while costs of urbanization may in fact rise as the city becomes larger, so do the benefits (as measured by income) but in an even larger proportion; and, as noted previously, the income increase shows no sign of abating even for the very largest cities.

It might, however, be noted that in at least one large city, namely London, a situation combining rising costs and static incomes has been observed (Eversley, 1972). Urban costs there are held to rise as a consequence of rising land prices, higher construction costs, accelerating obsolescence, and the difficulties of carrying out public works in old, congested centres together with the high fixed costs of running a large city. Falling incomes are the result of a drop in total population and the out-migration of higher proportions of middle and upper income groups. It is especially important to note that both population and employment in the larger British cities have been dropping over the 1960s and that the Greater London area now has some two million less inhabitants

Table 7.4 Size of city and income levels in Mexico

Size of City (Inhabitants)	Monthly family income (pesos)
Less than 2,500 (rural)	740
2,501–10,000	1,000
10,001–150,000	1,450
150,001–500,000	1,900
500,000 plus	2,800

than it would have had had the trends of the 1950s continued. The similar finding with regard to the United States is that, while there are signs of increasing per capita expenditure past some level of country size, per capita income rises four times faster within the same range, i.e. the marginal propensity to consume public services is about 25 percent (Alonso, 1968, pages 4 and 12). The large city is consequently seen as a vital part of the country's productive apparatus; not only is productive efficiency enhanced through spatial concentration, but the metropolitan region is characterized by a final demand which enables scale economies to be achieved, permits firms access to highly specialized input markets and institutions (labour markets, financial institutions, and public infrastructure and services), and linkages can be exploited. The large city therefore produces not only a wider array of goods and services than other places but, when duplication exists, it produces more cheaply than elsewhere (Wingo, 1970).

Before evaluating these findings it seems appropriate to note the growing distinction between development and welfare, or economic and social considerations, and the growing dissatisfaction with the use of the former alone. There is no inherent reason why any optimism should be narrowly economic in nature, and it could be related as easily to such yardsticks as accessibility health, crime and safety, or any other characteristic of the city which is measurable. This is precisely what is argued when indicators other than per capita income are used as proxies for benefits. A basic question is whether in fact such benefits can be classified into distinct economic and non-economic categories. Is an increase in longevity, for example, economic or non-economic? And yet for some types of project evaluation its quantification is indispensable (UNIDO, 1972). It may well be that 'economic' benefits are simply general benefits expressed in economic terms and that, therefore, there is no major distinction between one set of benefits and the other except the degree to which they are susceptible to economic measurement. Nevertheless it must be borne in mind that this is but one perspective from which to view the costs and benefits of urbanization even if the various proxies cannot neatly be aggregated into a single index due to the difficulties encountered in weighting the different components (Ternent, 1975, Appendix A).

Incomes are expected to be higher in the large cities for the following reasons:

1. Wage rates will be higher if the value of the marginal product of labour is higher in the larger city for reasons such as natural advantage, scale economies, agglomeration advantages, etc., although this situation will be transitory or a characteristic of a disequilibrium situation to the extent that such differences set in motion processes which tend to eliminate them, for example, migration from the lower to the higher wage area.

2. Population groups are not homogeneous and differences in per capita income may be due to education, age, sex, etc. rather than city size *per se*. However, if the larger cities attract the more skilled individuals, then higher per capita incomes will be observed there. This point is especially important when the difference between an optimal and an actual size of city is measured

in terms of a number of people for it will make a great difference if the people are young or old, skilled or unskilled, etc. (Wingo, 1972).

3. Compensatory payments may also account for differences in money incomes, for example, to live in an area with an unpleasant climate people have to be paid more than in a more-hospitable zone. Two empirical findings in the United States are that the cost of living rises with city size, and that non-monetary conditions become increasingly unfavourable as a city grows larger. If real income exclusive of non-monetary considerations is equal to money income deflated by city cost of living (i.e. money income is worth less where living costs are higher), and the equilibrium condition is that real incomes must be equated between cities, then money incomes must be higher in the larger city in order both to compensate for living costs there and the increasing disutility of the non-monetary cost (Hoch, 1972). In other words, firms must bribe workers through higher money wages to leave the smaller cities where psychic incomes are higher, or negative externalities are smaller (Wingo, 1972).

4. Finally, even in equilibrium, transport and migration costs will permit income differentials to be maintained.

Although there can be no quarrel with the documented fact that money incomes, even corrected for cost of living differences, are higher in the urban than in the rural areas, migration continues from the smaller to the larger cities in Latin America in spite of the high rates of unemployment evident in the latter together with the obvious disutilities of life in the large city. Furthermore the above argument rests in large part upon empirical observations, in particular the fact that the larger cities are more productive because the more productive firms are located there; that costs of living are higher in large than in small cities; and that wages and real incomes are higher in the large than in the small city. Before accepting the basic conclusion, namely that the benefits of urbanization, as measured by income received, are greater the larger the city, the above empirical propositions would have to be verified country by country in Latin America. The partial information for Brazil reported elsewhere suggests that real incomes were in fact not higher in the larger cities during the early 1960s and may well have been lower (Ternent, 1974). Furthermore, in at least one Brazilian city of considerable size, Recife, there has been no improvement in annual real per capita income over the 1960s.[11] The larger and reputedly more productive firms in Colombia were not typically concentrated in the larger cities as noted below. Studying a recent sample of 200 cities in Brazil, Boisier similarly comes to the conclusion that industrial productivity increases more than proportionately with the urban size up to a level of 200,000 inhabitants, but less than proportionately after 500,000 inhabitants (Boisier, 1973, page 107).[12]

The postulate that living costs rise with city size is based on rising congestion costs, increases in the prices of certain types of goods (notably housing and land), higher transportation costs due to longer distances travelled and so on. The empirical predictions noted above have been derived from these theoretical propositions. The first piece of empirical data which may be adduced in evidence

Table 7.5 Cost of living and size of city in the United States

City size	All Items	Housing	Transportation	Health and Recreation
3.5 million or more	26.3	29.4	25.4	28.4
1.4 to 3.5 million	25.2	28.2	22.5	27.2
250,000 to 1.4 million	24.6	29.3	18.7	26.5
50,000 to 250,000	23.4	27.0	17.4	26.4
25,000 to 50,000	22.6	25.8	17.5	24.2

Source: Monthly Labour Review, August 1972, p. 4.

is for the United States where the increase in the cost of living between December 1966 and December 1971 is shown in Table 7.5.

The basic conclusions to be derived from these data are that the increase in the overall index, as well as in all the subgroups (only some of which are shown), was greater in percentage terms in the larger urban classes than in the smallest; that acceleration in price increases was greater in the larger than in the smaller urban classes and that deceleration was slower in the former than the latter; and that, although the above changes appear to be small (ranging from 1.14 percent per quarter in the smallest size class to 1.33 percent in the largest), they are all statistically significant at the one percent level. The largest discrepancies in the data for the United States are for transportation, housing, health and recreation, all areas which are considered to be in crisis in the large cities.

Similarly, Indian data show price levels, to be higher in Bombay than in smaller cities (see Table 7.6). Again it can be seen that differences in cost of living are insufficient to offset money income differentials (assuming that in India they are of the same magnitude as in the United States and Mexico); the conclusion which is arrived at again is that real incomes are higher in the larger cities. Cost of living differences appear to be relatively small, on the order of ten percent, over a very substantial range of city size.

The above data may be contrasted with information for Colombia (see Table 7.7) which is presented in descending order of city size.[13] The most striking fact about these tabulations is the lack of consistency with expecta-

Table 7.6 Price level in selected Indian cities

City	Population		Price Index
Jalgaon	less than	100,000	96.79
Nanded	less than	100,000	89.38
Aurangabad		138,000	93.23
Sholapur		410,000	93.45
Poona		925,000	95.96
Nagpur		1,030,000	92.47
Bombay		5,100,000	100.00

Source: Chavan, 1971, p. 383; Davis, 1969.

Table 7.7 City size and increases in cost of living in Colombia 1968–1973

City	Population 1973 (thousands)	Cost of Living increases Workers	Cost of Living increases Employees
Bogotá	2,871	77.7	68.9
Medellín	1,410	83.6	73.5
Cali	926	88.8	82.9
Barranquilla	729	76.8	76.6
Bucaramanga	377	85.9	8.19
Manizales	204	89.3	84.7
Pasto	115	73.1	76.2

tions derived from the data for the United States and India presented above as well as from casual hearsay evidence. It must be pointed out, however, that the above data do not support the contention that Bogotá, the capital city of Colombia, is cheapest place to live; a strict interpretation is that the smallest increase in the cost of living among the cities for which there is evidence in Colombia over the period 1968–1973 was found in Bogotá for employees. For the worker index, Bogotá's increase was the smallest among the large cities (i.e. excluding Pasto). Even the suggestion of an inverse relationship between cost of living increases and city size in Colombia, and therefore accentuated money-income differentials, is surprising. However, any firm diagnosis would have to consider absolute price levels in the various cities.

The theoretical arguments proposed above rest implicitly on the conditions required to fulfil a full-employment equilibrium, which may or may not be appropriate in a developing country. The key assumption is that a worker receives the value of his marginal product which is quite unrealistic given the conditions observed in the cities of the developing countries. The other extreme, frequently posited in project evalutation, is that the social opportunity cost of labour, or the real cost to society of employing an additional worker, is zero. This is due to the fact that the cost of putting the person to work is what he would have produced elsewhere; being unemployed this cost is zero. But this may not be a realistic appraisal of the situation for the alternative to idleness is likely to be very unpleasant indeed as working conditions in the urban areas of many of the poorer countries are very bad. Some compensation is therefore required (UNIDO, 1972, pages 95–96). This is of course similar to the last reasons given above for differentials in wages and incomes for not only does migration involve a series of direct costs, but urban facilities also have to be provided for the migrant upon arrival at his destination.

The basic issue in the optimum city-size argument appears therefore to centre on the workings of the natural deconcentration mechanisms identified previously and the mobility of capital and labour to and from the largest or primate city. It would not appear possible to have too large a city if the market mechanisms function properly as externalities would be internalized, raising costs and squeezing high cost firms (and population) out to the smaller cities. The whole issue of excessive urban size appears to hinge on the behavioural premises adopted concerning the functioning of market mechanisms.

If they do not function, the primate city can grow too large as there is no diffusion of growth to other cities. One example of the inappropriate functioning of the market is put forward by Thompson (1972, pages 99–100), who hypothesizes that the benefits of urbanization will depend to a large extent upon the income level of the group in question. The welfare of the lower income group is postulated to rise rapidly up to moderately sized cities and then drop as diseconomies of scale set in, thus yielding a U-shaped benefit curve from the point of view of economy. The more educated and affluent classes, on the other hand seek variety, which rises with city size; this benefit curve may well increase indefinitely.[14] Excessive population concentration may be defined, therefore, in one of two ways: fewer low income or less educated households prefer living in the large city than are needed there in relation to the number of higher income and more educated households which prefer to live there (i.e. more lower income households are 'forced' to live in the large city than would do so of their own volition); or conversely too few of the higher income group are willing to live in the smaller urban areas and provide employment for all the lower income households which would prefer to live there. This bias in favour of the large city is reinforced by the fact that those who wish to live there typically have a great deal to say about industrial location. Questionnaire data show that 'personal reasons' are prominent among the location factors identified (although this covers a great deal more ground than 'variety') (Hoover, 1971, pages 61–62).

Another important reason explaining the presence of excessive concentration relates precisely to the income considerations analysed above, that is that the great mass of skilled and semi-skilled workers have given up whatever influence they might have had on work, and consequently residential, location through the imposition by unions of a spatially invariant money wage, again frustrating the operation of the natural market mechanisms which would have reinforced any decentralization tendencies present. The premise here is that labour unions suffer from a money illusion in that it is the money wage which is of interest rather than its purchasing power, or the real wage deflated by the cost of living in particular cities. In addition average rather than marginal costs are charged for public services of all sorts.

The suggestion implicit in the above paragraphs is that the benefits of urbanization, at least as measured by money and real incomes, are not necessarily higher in the larger Latin American cities, and this is borne out by empirical research into the relationship between average urbanization and a variety of measures of wellbeing.

Another element in the optimum-size controversy concerns the relationship between public costs and city size. And while urban public service costs are only one element in the determination of optimum size and are a relatively small component of total costs this should not obscure the possibility that they may be a crucial element. Thus, if construction costs rise with higher densities and larger cities, while land costs decline (although land cost is held to be a transfer payment and consequently of no importance in the investment decision), infrastructure costs which have a U-shaped form may very well be the critical determinant. Similarly construction costs may well be a small proportion of the total economic product of a city, and differences in discounted present values of construction between cities may be smaller still, but if they are variable to a greater degree than other considerations in the investment decision, they may well assume an importance out of all proportion to their actual size. Finally, it is held that managerial diseconomies may be the fixed factor causing the onset of diminishing returns to city size. Impressionistic evidence suggests that, on the contrary, in Latin America it is the small authority that is inefficient, although exactly how efficiency is defined is not clear. It should be noted, however, that even with the obvious concentration of urban managerial talent in the larger cities, it is precisely these which are unable to cope with the demands made on them. The blanket statement that small authorities are inefficient also ignores the fact that there are a series of functions, mainly those which are not subject to economies of scale, which are most appropriately carried out by precisely these smallest levels of government. The inefficiency of a particular level of government must also be evaluated within the particular context in which it operates. Thus, in Latin America much is made of the fact that municipal power has been eroded over time with more senior levels assuming its functions.

It would appear that the three most damaging points made against the optimum city size concept are: that the level of knowledge concerning individual firms is so low as to preclude any legitimate conclusions; that any optimum arrived at for firms would not necessarily be optimal from the point of view of households; and that each public service may also have an optimum which will not necessarily coincide with either of the above (von Boventer, 1973). Furthermore, any such optimum must be dynamic and take into account the initial conditions as well as the speed of development. In spite of these difficulties, an empirical regularity known as the rank-size rule has been found to apply in many instances. A major contradiction is apparently present here for it would not seem possible to simultaneously hold that there are urban hierarchies characterized by the lodging of higher level functions in the larger cities and that there is one city size which is optimal. Furthermore, as an urban area proceeds up the city-size scale it acquires new and different attributes according to central place theory. Comparisons of small and large cities, therefore, may be quite hazardous (Richardson, 1972, page 34). These would appear to be such serious drawbacks, that one would be justified in abandoning the unique optimum city size concept completely. In passing it may be noted that the rank-size rule is considered too crude a measure of urban policy and the recommended course when deviations from it are present is to look for specific causes. The rank-size rule is not an ideal state and urban policy should not attempt to correct deviations from it.

If there is no unique optimum city size, it may be postulated that each city has an optimum size depending upon available technology and the production function which is adopted (Hoch, 1972, page 300). An extension of this argument is to consider a family of benefit curves which rise as more functions are added to the city; in other words an optimum for each rank order is envisaged (Richardson, 1972, page 35). The argument here is that within a given context there is such a thing as an optimum size for a market town, for a regional capital, and so on, and to the extent that it is about to be overcome a new one should be established. It is reasonable to suppose that there will be a range of sizes which will qualify and that it would be most appropriate within each rank to choose cities at the lower end of the range for, if some of the larger ones were to be chosen, they would rapidly enter the diminishing returns stage (Thompson, 1972, page 115). The assumption here is again that some sort of an optimum really does exist although whether it is an absolute size or a rate of growth becomes of considerable interest (it should be borne in mind that the two are of course intimately connected).

A further alternative to the optimum city size is a settlement pattern such that, given economic activity and consumer preferences, no firm or worker wants to move from his present location. At the margin no dollar of investment or worker could be reallocated so as to improve welfare, where this is defined as a compound of economic and psychic incomes (Wingo, 1972, page 18). Clearly a further condition would have to be met here and that is that not only would no economic unit wish to move, but the society itself would not want such a move to take place—every unit could be in its preferred location, but on the basis of social cost-benefit analysis, it might be desirable to provoke internal migration and rearrangement of the settlement pattern.

This last point leads to a discussion of what appears to be a most important myth in urban studies at the present time and that is the interpretation given to the phrase 'the irreversibility of urbanization'. In a trivial sense urbanization is obviously irreversible; as development proceeds everything seems to point to the fact that an increasing proportion of the national population will live in cities. While irreversible, such a transition is also finite, for not more than 100 percent of a nation's population can inhabit its cities. The most obvious development then is an asymptotic approach to the upper bound. However, urbanization is also presented as a process which, under the circumstances in Latin America, and perhaps much of the developing world, culminates in a movement to the primate city. Its growth is thus reinforced until rates so high have been reached that observers begin to foresee cities of 50 million inhabitants as a distinct possibility for Latin America in the not too distant future. This is a variant on the irreversibility and tidal wave nature of urbanization and should be discarded. The entire burden of the above discussion has been that concentration is reversible and that excessive concentration is due in part to the working of the system. If public policy has any effect at all, it will have an impact on the workings of the system and should make it possible to counteract the forces which up to the present time have made excessive concentration inevitable. One immediate example is the payment of higher public salaries in the large than in the small cities. Another is the maintenance of overvalued exchange rates.

The conclusion that should be arrived at is that, if absolute size in fact cannot be influenced too much, growth rates are subject to influences from major events in the life of the country and in public policy. Recent data for Brazil support this proposition. A major influence on urbanization has been the shift in agricultural output in the State of São Paulo from labour intensive coffee production to land and capital intensive cotton and cattle. Rates of urbanization have, under these conditions, varied most extraordinarily between the decade of the 1950s to that of the 1960s suggesting that under the conditions of very rapid urbanization, substantial changes can be induced in the national system of cities quite rapidly. It should be pointed out that there is no need for there to be an 'optimal' growth rate for growth to be too fast or too slow to be properly assimilated. Although in principle it would appear that an urban area would be well advised to aim at the average national growth rate as its immediate target (that is zero net migration), where demand-pull growth is evident (or output growth requires additional population and labour) a higher than average growth rate would be fully justified. Alternatively if an underemployment push is driving people from the countryside to the primate city, then a different policy mix (combining measures to provide employment in agriculture), the development of small centres of rural-urban integration, and the strengthening of alternatives to the primate city is in order (Thompson, 1972, page 116). The policy of diverting migrants to cities other than the largest one will have to be pursued with a considerable amount of care, however, for it is very easy to overwhelm them. Only a small fraction of the increment to a large city's population could easily double the population of a smaller town. It then becomes necessary to choose a subset of the smaller cities such that the absolute decrease in the large city's growth will represent a manageable increment to the population of each of the smaller cities.

Opponents of the optimum city size concept also ask if the spatial distribution of the population within the larger cities may not be as important as their actual size. In other words, may it not be that the increasing costs both monetary and non-monetary which are said to characterize the large city are due as much to the imperfect organization of the city as to its excessive size. External diseconomies may as easily be a function of density as size, for in traditional fashion city growth may be conceived of as proceeding at both the extensive and the intensive margin. Thus, when to the policy of affecting the growth rate of large cities and the alternatives to them is added the stated policy objective of improving the spatial organization of the former, the conclusions arrived at are in fact identical to those derived from the simpler static optimum city size model. It should be noted that as external diseconomies are as much a function of densities as of absolute size, it is possible to derive a U-shaped cost function as densities rise. This close correspondence between size and densities has still to be demonstrated empirically for Latin America, but to the extent that it does exist, the focus on size should be complemented immediately by a focus on space, and the conclusions of the simpler model are borne out. The operational conclusions appear to be as follows.

1. To the extnet that something like diminishing returns does in fact exist, migration flows should be diverted to other, smaller cities.
2. Target cities for such a policy of deconcentration should fall towards the lower end of the range of middle-sized 'efficient' cities (i.e. somewhere around 100,000 inhabitants rather than one million; smaller cities are vulnerable to footloose industry which flees higher taxes or stricter pollution controls while individual governments in a large metropolitan area are rather similar constraints).
3. In order to avoid overwhelming any one middle-sized city, it is necessary to enhance the attraction of that subset of intermediate cities such that the growth of any one of them is not unmanageable, but that makes a significant impact all told on the growth rate of the largest city.
4. Although this has not been included in the body of the paper above, there is a presumption that the efficiency of an urban network is related not only to the internal functioning of a city and the distribution of functions between cities, but to distances between cities. From this is deduced the general principle that alternatives close to the large city which one wishes to decentralize should be small high quality cities which will not complete with the primate one but rather complement it; the further away one gets from the primate city, the greater the need to achieve scale economies as rapidly as possible and therefore one should concentrate on larger cities.

Application of a framework similar to the above in one Latin American country, Brazil, yielded very heartening results. As opposed to the findings for the United States reported above, Tolosa (1973, page 640) found that the benefits of urbanization, as measured by productivity, wages and family incomes, rose by 50 to 100 percent from the smaller to the largest cities, whereas the per capita expenditures rose from 3–4 times where economic infrastructure and seven times where social infrastructure was concerned. Where infrastructure is concerned the data are quite recent, being for the year 1969. In spite of the well-known drawbacks and difficulties associated with the use of expenditures to measure urbanization costs, the magnitude of the increases is sufficient to warrant adopting a dispersal policy of strengthening the intermediate size cities which have already had an important role to play in the recent development of the Brazilian urban system.

NOTES

1. 'Entropy (derived from the second law of thermodynamics) is a measure of the degree of equalization reached within a system, and is maximized when the system reaches equilibrium' (Richardson, 1973, pages 147–148).
2. The measure of primacy used by El-Shakhs consists, for a particular city, of the average of the ratios of the sizes of all cities smaller than itself to its own size. Primacy for the entire system of cities is defined as the average of the values for each of the cities within the system (El-Shakhs, 1972).
3. One indication of the unsatisfactory nature of the indicator of development is that the 'developed group' included countries with incomes ranging in 1969 from over 4,000 dollars per capita to less than 200 (World Bank, 1971). Furthermore, the

primacy index ranks countries in Latin America quite differently from the ways in which conventional wisdom or prior analysis do. The Latin American countries which were identified as most primate by all other methods (Argentina, Chile, Venezuela, and Mexico although there was insufficient data for the most primate country of them all, Uruguay) are precisely those considered least primate by El-Shakhs. Further experimentation was justified for there seemed no particularly good reason to prefer other indicators a priori to the one he used. While El-Shakhs explicitly wishes to avoid distorting the effects of a few deviations within the urban system, the analysis of primacy seems to focus precisely upon such deviations at the national (Lima) or regional (Guayaquil) levels. The Arriaga measure mentioned above also tends to eliminate extremes through its averaging procedure.
4. See UNRISD (1970), Davis (1969), and IDB (1972), all supplemented by recourse to census data where needed. The Davis measure of primacy is the population in the largest centre divided by the sum of the population in the next three largest cities.
5. If Y is the four-city index of first city primacy, and X is the UNRISD index of development, the least squares regression was:
$Y = 3.45 - 2.286 X \qquad R^2 = 0.167$
6. McGreevey's use of the chi square statistic accords with the intuitive reasoning that primacy for a system of cities should be measureable in terms of the deviations of observed variables from their expected values as calculated according to some acceptable rule, in this case the lognormal or rank-size distribution. For an array of cities the lognormal distribution enables the size of any one city to be estimated as a proportion of the total population in the entire set of cities. Multiplication of the lognormal factor by the total population in the array of cities yields the expected population of the city. The procedure then is to estimate the squares of the deviations of the observed from the expected values to calculate the chi square statistic for each array. As deviations are found at the upper end of the array, deviations from the expected values can be taken as indicators of primacy.
7. The preferred index, on the basis of its correlation with a set of development indicators, is $R/i - r$, where R is the proportion of a country's population resident in cities with at least 100,000 inhabitants, and r is the proportion in cities with more than 20,000 (Subramanian, 1971). The data for Latin America, the United States being included solely for the purpose of comparison, were as follows:

	Primacy	Development Index
United States	1.228	111
Argentina	.931	73
Chile	.658	61
Venezuela	.525	63
Mexico	.362	44
Colombia	.325	46
Brasil	.271	38
Ecuador	.236	31
Peru	.225	37
D. Republic	.128	30

8. See Friedmann (1968). On the other hand, Boulding (1970) argues that the real measure of well-being is not income but the state or condition of the person or society.
9. Berry, page 152. For literature suggesting other early disagreement with this position see Richardson (1972).
10. Such as the living conditions of the urban poor, the obvious disadvantages including congestion, inadequate infrastructure and pollution, and the ferment, chaos and agitation for political reform found there (Mills, 1971; Harris, 1971).
11. From a base level of 440 *cruzeiros* in 1960 (at 1971 prices), income rose to 710 *cruzeiros* on average during 1961–1962, only to fall to 560 *cruzeiros* in 1970 and 420 *cruzeiros*

in 1968 (Cavalcanti, 1972, page 88). See Chapter 5, pages 125–135, for more detailed comments on the Brazilian case.
12. It might be noted that in neither of these cases was city size corrected for industrial mix. In another study (Rocca, 1970) found that higher levels of labour productivity by sector were generally associated with the larger concentrations of manufacturing activity even after allowing for invested capital per worker and size of plant.
13. Cost of living data from Banco de la República, 1973 while population data is from DANE, 1974. Percentage change in cost of living for 1973 from DANE communique (published in El Espectador, Bogotá, January 7, 1974, page 1) applied to average 1972 level of cost of living index in each city to arrive at increase for whole period.
14. Variety need not however solely be a function of city size, but can vary widely among cities of a given size as a result of technological change and conscious policies designed to foster variety (Hoover, 1971, page 382).

REFERENCES

Alonso, W. (1968) 'Urban and regional imbalances in economic development', *Economic Development and Cultural Change*, **17**, 1–14.
Arriaga, E. E. (1970) 'A new approach to the measurement of urbanization' *Economic Development and Cultural Change*, **18**, 206–218.
Banco de la República, (1973) *Revista del Banco de la República*, **46**.
Berry, B. J. L. (1964) 'City-size distributions and economic development', in Friedmann, J. and Alonso, W. (Eds.) *Regional development and planning*, Cambridge, Mass.: M.I.T. Press, 138–152.
Berry, B. J. L. (1971) *Urban hierarchies and spatial organization in developing countries*, (paper presented to the Rehovot Conference on Urbanization and Development in Developing Countries, Israel, 16–24 August, 1971).
Beyer, G. H. (Ed.) (1967) *The urban explosion in Latin America; a continent in the process of modernization* (Ithaca, New York: Cornell University Press).
Boisier, S. (1972) 'Comentarios', *Revista Latinoamericana de Estudios Urbano Rurales (EURE)*, **2**, 117–120.
Boisier, S. (1973) 'Localización, tamaño urbano y productividad industrial: un caso de estudio de Brasil', *Revista Interamericana de Planificación*, **6**, 87–112.
Boulding, K. E. (1970) 'Fun and games with the gross national product—the role of misleading indicators in social policy', in Helfrich, Jr. H. H. (Ed.), *The Environmental Crisis* (New Haven: Yale University Press), 157–175.
Cavalcanti, C. de V. (1972) 'A renda familiar e por habitante na cidade do Recife', *Pesquiza e Planejamiento Económico*, **2**, 81–104.
Centro Interamericano de Vivienda y Planeamiento (CINVA), (1970) *Boletín Informativo*, January, 5–6.
Chavan, B. W. (1971) 'Purchasing power of a rupee in different cities', *Indian Economic Journal*, **18**, 377–385.
Culbertson, J. M. (1971) *Economic development: an ecological approach* (New York, Knopf).
Davis, K. (1969) *World urbanization 1950–1970, Volume I: Data for cities, countries and regions*, (Berkeley, California: University of California).
Departmento Administrativo Nacional de Estadística (1974) *Tabulados del Censo de 1973*, Bogotá: (DANE).
El-Shakhs, S. (1972) 'Development, primacy and systems of cities', *Journal of Developing Areas*, **7**, 11–36.
Eversley, D. E. C. (1972) 'Rising costs and static incomes: some economic consequences of regional planning in London', *Urban Studies*, **9**, 347–368.

Friedmann, J. (1968) 'The strategy of deliberate urbanization', *Journal of the American Institute of Planners*, **34**, 364–373.
Hahn-Been, L. (1968) *Introductory presentation of conference objectives in the new urban debate*, report of the Pacific Conference on Urban growth, Honolulu, Hawaii, May 1–12, 1967 (Washington, D.C.: Agency for International Development).
Harris, J. R. (1971) 'Urban and industrial concentration in developing economies: an analytical framework', *Regional Urban Economics*, **1**, 139–152.
Hirsch, W. Z. (1973) *Urban economic analysis* (New York: McGraw-Hill).
Hoch, I (1972) 'Income and city size', *Urban studies*, **9**, 299–328.
Hoover, E. M. (1971) *An introduction to regional economics* (New York: Knopf.)
Inter-American Development Bank (1972) *Urban population growth series* (Washington, D.C. mimeo.), various volumes.
International Bank for Reconstruction and Development (IBRD) (1971) *Trends in developing countries* (Washington, D.C.)
International Bank for Reconstruction and Development (IBRD) (1972) *Economic growth of Colombia: problems and prospects* (Baltimore and London: Johns Hopkins University Press).
Instituto Brasileiro de Geografia y Estatística (1971) VII Rencenseamento Geral, 1970: *Resultados preliminares tabulacoes avancadas do Censo Demográfico* (Rio de Janeiro: IBGE).
Johnson, E. A. J. (1970) *The organization of space in developing countries* (Cambridge, Mass.: Harvard University Press).
Kuznets, S. (1966) *Modern economic growth; rate, structure and spread* (New Haven: Yale University Press).
Linsky, A. S. (1965) 'Some generalizations concerning primate cities' in Breese G. (Ed.) (1969) *The city in newly developing countries*, (Englewood Cliffs, N. J.: Prentice-Hall), 285–294.
McGreevey, W. P. (1971) 'A statistical analysis of primacy and lognormality of Latin American Cities, 1750–1920', in Richard Morse and coworkers (Eds.) *The urban development of Latin America, 1750–1920*, (Stanford, California: Stanford University Press), 116–129
Mehta, S. K. (1964) 'Some demographic and economic correlates of primate cities: a case for revaluation', in Gerald Breese (Ed.) (1969) *The city in newly developing countries* (Englewood Cliffs, N. J.: Prentice-Hall), 295–308.
Mera, K. (1973) 'On the urban agglomeration and economic efficiency', *Economic Development and Cultural Change*, **21**, 309–324.
Miller, J. and Gakenheimer, R. (1969) 'Editors' preface', *American Behavioural Scientist*, **12**, 48.
Mills, E. S. (1971) *City sizes in developing countries* (paper presented to the Rehovot Conference on Urbanization and Development in Developing Countries, Israel, 16–24 August 1971).
Neutze, G. M. (1965) *Economic policy and the size of cities* (Canberra: Australian National University Press).
Rehovot Conference on Urbanization and Development in Developing Countries (n.d.) *Interim Report* (Israel, August 16–24, 1971).
Revista Mexicana de Ingeniería y Arquitectura, (1973) **52**.
Richardson, H. W. (1969) *Regional economics* (New York: Praeger).
Richardson, H. W. (1971) *Urban economics* (Harmondsworth, England: Penguin).
Richardson, H. W. (1972) 'Optimality in city size, systems of cities and urban policy: a sceptic's view', *Urban Studies*, **9**, 29–48.
Richardson, H. W. (1973) *The economics of urban size* (Westmead and Lexington: Saxon House and Lexington).
Rocca, C. A. (1970) 'Productivity in Brazilian manufacturing', in Bergsmann, J. (1970) *Brazil: industrialization and trade policies* (London: Oxford. U.P.), 222–241.

Sovani, N. (1964) 'The analysis of "overurbanization"', in G. Breese (Ed.) (1969) *The city in newly developing countries* (Englewood Cliffs, N. J.: Prentice-Hall).
Stöhr, W. B. (1972) *El desarrollo regional en América Latina; experiencias y perspectivas* (Buenos Aires: Ediciones SIAP).
Stvan, J. (1973) *Physical, socio-economic, and environmental planning in countries of Eastern Europe* (Stockholm: National Swedish Building Council).
Subramanian, M. (1971) 'An operational measure of urban concentration', *Economic Development and Cultural Change*, **20**, 105–116.
Ternent, J. A. S. (1970) 'Algunas consideraciones económicas sobre una política urbana para Colombia' in Universidad Nacional de Colombia, *Documentos Técnicos* No. 2 (Bogotá: Centro de Investigaciones para el Desarrollo), 18–127.
Ternent, J. A. S. (1975) 'Hacia políticas nacionales de urbanización en América Latina', *América Latina: distribución espacial de la población*, Ed. R. Cardona (Bogotá, Corporación Centro Regional de Población), 321–424.
Tolosa, H. C. (1973) 'Macroeconomia da urbanização brasileira', *Pesquisa e Planejamento Econômico*, **3**, 585–644.
Thompson, W. R. (1972) 'The national system of cities as an object of public policy', *Urban Studies*, **9**, 99–116.
United Nations (1968) 'Urbanization: development policies and planning', *International Social Development Review*, **1** (New York: United Nations).
United Nations (1970) *Urbanization in the United Nations Second Development Decade* (New York: United Nations).
United Nations (1971) *Human Settlements*, **1**.
United Nations (1972) *Report of the inter-regional seminar on the financing of housing and urban development*, (New York: United Nations).
United Nations Industrial Development Organization (UNIDO) (1972) *Guidelines for project evaluation* (New York: United Nations).
United Nations Research Institute for Social Development (1970) (UNRISD) *Content and measurement of socio-economic development: an empirical enquiry* (Geneva: UNRISD).
Urdaneta, A. (1973) 'Costos del desarrollo urbano', *Revista Interamericana de Planificación*, **7**, 80–100
Von Boventer, E. (1971) *Urban hierarchies and spatial organization in developing countries* (paper presented to the Rehovot Conference on Development and Urbanization in Developing Countries, Israel, 16–24 August 1971), 16.
Von Boventer, E. (1973) 'City-size systems: theoretical issues, empirical regularities and planning guides', *Urban Studies*, **10**, 145–162.
Williamson, J. G. (1965) 'Regional inequality and the Process of national development: a description of the patterns', in L. Needleman (Ed.) (1968) *Regional Analysis* (Harmondsworth, England: Penguin), 99–158.
Wingo, L. (1971) *National objectives, development and metropolitan concentration* (paper presented to the Rehovot Conference on Urbanization and Development in Developing Countries, Israel, 16–24, August 1971), 9.
Wingo, L. (1972) 'Issues in a national urbanization policy for the United States', *Urban Studies*, **9**, 3–28.

Author Index

Abiodun, J. O., 87
Abrams, C., 69
Achebe, C., 11
Acosta, M., 58
Adams, D. W., 64
Adelman, I., 118
Agarwala, A. M., 165
Alers, O., 60
Allen, K., 2
Almeida Andrade, T., 127, 140
Alonso, W., 3, 5, 8, 123, 125, 137, 180, 183
Andrews, H. F., 86
Appalraju, J., v, 16
Applebaum, R. P., 60
Argentina, 140
Arriaga, E. E., 172, 177, 182, 192
Arrighi, G., 165
Atkinson, A. B., 124

Baer, W., 114
Balán, J., 5, 52, 54, 62, 63, 66
Banco Central de Reserva del Perú, 141
Banco de Bilbao, Spain, 141
Banco de la República, Colombia, 193
Banco do Nordeste do Brasil, 131–2
Banks, J. A., 57
Bantan, M., 57, 58
Barber, W. J., 89
Barboza de Araujo, A., 126
Barlow Commission, 2
Baster, N., 138
Bauer, P. T., 6, 12
Bazán, C., 58
Beckinsale, R. P., 110
Bergsmann, J., 5, 194
Bernard, P., 2

Berry, B. J. L., 3, 86, 105, 107, 118, 149, 173, 178, 180, 192
Berry, L., 110
Beyer, G. H., 171
Birmingham, W., 141
Blaikie, P. M., v, 15, 25, 28, 37
Blaug, M., 165
Bobek, H., 88
Bogue, D. J., 59, 62
Boisier, S., 6, 140, 178, 184
Bolivia, 140
Borts, G. H., 120
Boudeville, J. R., 144, 149
Boulding, K. E., 172
Bowen, I., 122
Brandt, H., 96
Bray, J., 104
Brazil, 140
Breese, G., 47, 69, 72, 194
Brigg, P., 51, 72
Brookfield, H., 5, 16
Browning, H. L., 5, 52, 53, 54, 62, 63, 66
Bruner, E. M., 60
Bugnicourt, J., 141
Butterworth, D., 72
Byerlee, D., 72

Caldwell, J. C., 51, 52, 60, 62, 71
Calvacanti, C., 128, 132, 193
Calvacanti, R., 125, 129
Cameron, G., 165
Cardona, R., 5, 52, 55, 61, 63, 66, 67, 68, 72
Cardoso, F. H., 5, 16
Carrillo Arronte, R., 141
Carvalho, J. A. M., 135

Chavan, B. W., 185
Chenery, H. B., 13
Chile, 140
Chisholm, M., 2
Chorley, R. J., 105
Christaller, W., 86, 89
CINVA, Colombia, 178
Clark, C., 13
Clay, E. J., 29, 32
Cleaver, D., 37
Clout, H. D., 2
Colombia, DANE, 193
Commoner, B., 121
Connell, J., 5, 84
Conning, A., 52, 58
Converse, J. W., 74
Coraggio, J. L., 13
Cornelius, W., 69
Cotler, J., 5
Coutinho, A. B., 134
Culbertson, J. M., 171
Cummings, H., 60, 61
Currie, L. L., 13

Daland, R. T., 11
da Mata, M., 129
Darwent, D. F., 145
Davies, W. K. D., 87
Davis, K., 13, 48, 56, 174, 175, 185, 192
Daza Roa, A., 116, 141
de Graft-Johnson, K. T., 60, 67
de Oliviera, O., 72
De Voretz, D. J., 50
Deshmukh, M. B., 74
Dickenson, J. P., 125
Dorner, P., 67
Dumont, L., 35
Dumont, R., 7
Dwyer, D. J., 5

Eames, E., 52, 60, 72
Easterlin, R. A., 120
EFTA, 145
Eicher, C. K., 72
Elizaga, J., 51
Elliot, C., 116, 122
El-Shakhs, S., 173, 174, 175, 177, 180, 191, 192
Emmanuel, A., 99

Enos, J. L., 6
Estall, R., 2
Eversley, D. E. C., 182

Faletto, E., 5, 16
Fallon, K., 165
Feindt, W., 53, 62, 63
Fergus, M., 85
Fishlow, A., 124, 126
Fliegel, F., 35
Flinn, W. L., 67
Ford Foundation, 6, 144
Frank, A. G., 81, 89, 99, 121
Frankel, F. R., 37
Fiedlander, D., 56
Friedmann, J., 3, 4, 5, 11, 116, 118, 122, 149, 192
Fry, A. J., 94
Funnell, D. C., v, 15, 87, 91, 94, 97, 106
Furtado, C., 16, 114

Gakenheimer, R. A., 6, 171
Galbraith, K., 123
Gaur, R. S., 71
Gauthier, H. L., 127
Geisse, G., 6, 13
Gerken, E., 96
Ghana, 141
Gilbert, A. G., v–vi, 4, 5, 7, 16, 53, 60, 65, 72, 113, 122, 129, 150
Glentworth, G., 84
Godfrey, E. M., 50
González Casanova, P., 5, 121
Good, C. M., 83, 97
Goodman, D. E., vi, 8, 16, 114, 125, 129, 133
Gordon, J. E., 134
Goshi, T., 98
Gould, P. R., 5, 90, 150, 153
Gould, W. T. S., 74, 93
Graham, D. H., 128
Gravier, J. F., 2
Griffin, K., 6
Grove, D., 86, 87
GTDN, Brazil, 126, 127
Gugler, J., 65, 66, 67, 72
Gutkind, P. C. W., 81

Hägerstrand, T., 165
Haggett, P., 105
Hahn-Been, L., 169, 171
Hall, P., 2, 9
Hall, S., 93
Hance, W. A., 72
Hansen, N. M., 2, 165
Hanson, A. H., 7
Hardoy, J. E., 6, 58
Harris, J. R., 50, 192
Harriss, B., 33, 34
Harvey, M. E., 106
Hauser, P. M., 75
Hawkins, H. C. G., 98, 99
Heer, D. M., 39
Helfrich, H. H., 193
Hermansen, T., 145, 149
Herrick, B., 52, 58, 61, 63
Hertz, H., 13
Hilhorst, J. G., 5
Hill, P., 83
Hirsch, W. Z., 197
Hirst, M. A., 80, 85, 102, 103
Hoch, I., 182, 188
Holland, S. K., 9
Holmans, A. E., 2
Hoover, E. M., 187
Hoselitz, B. F., 89
Hoskins, M., 32
Houston, J. M., 110
Hoyle, B. S., 5, 110
Hume, I. M., 129, 133
Huszar, L., 86, 87

IBRD, 69, 118, 134, 191
IBGE, Brazil, 171
IDB, 171, 192
IDRC, Ottawa, 51, 60, 66, 67
Illich, I., 12, 121
ILO, 114, 185
India, 34, 141, 154
Intermet, 51, 60, 66, 67
Iran, 160, 164
Isard, W., 149

Jacobson, L., 138
Jansen, C. J., 59
Jelin, E., 5, 52, 54, 66
Jensen, R. C., 116—117

Johnson, E. A. J., 5, 37, 96, 150
Johnson, G. E., 71, 178

Kabwegyere, T. B., 95
Kade, G., 83, 87, 88
Kamalamo, P., 87
Keeble, D. E., 5, 116, 150
Kenya, 141, 160, 161—163
Kim, Y., 60, 61
King, M., 26, 92
Kivlin, J. E., 35
Knodel, J., 52
Koenigsberger, O. H., 153
Kosiński, L., 72
Kuklinski, A., 145
Kuznets, S., 13, 115, 169

Ladejinsky, W., 24, 29
Lal, D., 7
Langlands, B. W., 83, 93, 109
Langoni, C. G., 124, 126, 129, 130
Laquian, A. A., 72
Larrimore, A. E., 143
Lasuén, J. R., 5, 113, 149
Lee, F. S., 60, 65
Leeds, A., 54, 67
Leeds, E., 54, 67
Leinbach, T. R., 90
Lerner, 13
Leys, C., 6, 7
Lichfield, N., 85
Linsky, A. S., 173
Lipton, M., 7
Livingstone, I., 102, 104
Lösch, A., 86
Lowder, S., 60, 65
Lundqvist, J., 102, 103—104
Lutz, V., 9

Mabogunje, A. L., 57
Mabury, M. A., 84
McCallum, J. D., 12
McGee, T. G., 72
McGlashan, N. D., 92
McGreevey, W., 64, 176, 192
Mackay, J. K., 98
Mackenzie, M. K., 96
Maclennan, M. C., 2
McMaster, D. N., 81

Maimbo, F. T., 94
Malaysia, 154, 155
Malisz, B., 85
Mangin, W., 51
Manners, G., 2
Marabelli, F., 116, 141
Marris, P., 6, 83, 95
Marx, K., 36
Mascarenas, A. C., 87
May, D. A., 39
Mayfield, R., 35
Meadows, D. H., 121
Mehta, S. K., 173
Mera, K., 182
Metwally, M. M., 116–117
Mexico, 141
Middleton, B. J., 77, 107
Miller, J. P., 6, 171
Mills, E. S., 173, 192
Miner, H., 108
Misra, R. P., 5, 145
Moore, B., 24
Morris, C. T., 118
Morrison, P. A., 68, 70, 71
Morse, R. M., 194
Moseley, M. J., 108, 144, 145, 146, 147, 164
Mukasa-Kintu, 93
Myrdal, G., 12, 120–121, 122, 149

Nepal, G. S., 71
Neustradt, I., 141
Neutze, G. M., 179
Nicholson, M., 121
Nove, A., 108

Obudho, R. A., 87
Ocitti, J. P., 85
O'Connor, A. M., 5, 81
Odell, P. R., 5, 118, 122
Odingo, R. S., 141
Okulo-Epak, F., 85
Okun, B., 120
Omaboe, E. N., 141
Ominde, S. H., 141
Oshima, H. T., 139

Pakistan, 154
Pakkasem, P., 141

Parr, J. B., 107, 117, 124
Pedersen, P. O., 105
Pellerin, G., 133
Perroux, F., 143, 144, 148, 165
Pescatello, A., 76
Pioro, Z., 166
Poffenberger, T., 40
Ponzio, M., 87
Portais, M., 93, 96, 99
Prachuabmoh, V., 52
Prakash, V., 138
Pred, A., 86
Presad, K., 24
Preston, D. A., 5, 118
Prothero, R. M., 72
Pryor, R. J., 73
Pye, L. W., 74

Ramírez, R., 13
Rao, V. S. L. P., 5, 145
Ravenstein, 60
Redcliffe-Maud Commission, 89
Rehovot Settlement Study Centre, 52
Rehovot Conference, 172
Reiner, T. A., 122
Richardson, H. W., 8, 179, 182, 188, 189, 191, 192
Richardson, R. W., 120
Riddell, J. B., 5, 150
Roberts, B. R., 5
Robock, S. H., 114, 125
Rocca, C., 193
Rodwin, L., 2, 3
Roett, R. J. A., 114, 138
Roneche, K., 98
Rosenstein-Rodan, P. N., 148
Rostow, W. W., 149
Roy, P., 35
Rweyamamu, J., 166

Safier, M., vi, 16, 80, 85, 153
Sahota, G. S., 51
Salm, C., 133,
Saylor, R. G., 102, 104
Schubert, B., 96
Schultz, T. P., 50, 57
Schumacher, E. F., 121
Schumpeter, J. A., 167
Scitovsky, T., 149

Scott, A. J., 80, 90
Scrimshaw, N. S., 134
Seers, D., 6, 7
Seidman, A., 103
Semple, R. K., 127
Sen, A. K., 117, 124
Sen, L., 35
Shanin, T., 36
Shelty, N. S., 35
Sicat, G. P., 141
Simmons, A. B., vi, 5, 15, 52, 61, 63, 66, 67, 68, 72
Singh, K. P., 35
Singh, S. P., 165
Smith, J. A., 110
Smith, M. L., 60, 106
Soja, E. W., 5, 150
Somerset, A., 83, 95
Southall, 107
Sovani, N., 173, 181
Splansky, J. B., 87
Stanford Research Institute, 69
Stavenhagen, R., 5
Stein, J. L., 120
Stern, C., 72
Stöhr, W. B., 4, 5, 122, 170
Streeten, P., 7
Stvan, J., 170
Subramanian, M., 192
SUDENE, Brazil, 126
Sundaram, K. V., 5, 145
Sundrum, R. M., 114
Sunkel, O., 121
Suzuki, T., 107

Tanzania, 141, 155, 160, 163–164
Taylor, D. R. F., 5, 80, 86, 87, 88
Ternent, J. A. S., vii, 16, 171, 172, 180, 182, 183, 184
Textor, R. B., 65
Thailand, NEDB, 141
Thoman, R. S., 140
Thomas, I. D., 93
Thomas, M. D., 167
Thompson, W. R., 178, 187, 189, 190

Tinkler, K. I., 106
Tissandier, J., 85
Tiwari, S. G., 141
Todaro, M. P., 50
Todd, D., 145
Tolosa, H. C., 191
Törnqvist, G., 167
Todaro, M. P., 50
Travieso, F., 5
Trinidad and Tobago, 154, 155
Trudgill, P., 35
Turner, J. F. C., 54

Uganda, 84
Unikel, L., 141
UN, 169, 170, 172
UNECAFE, 114
UNECLA, 140
UNIDO, 183, 186
UNRISD, 113, 145, 174, 175
Urdaneta, A., 195
Utria, R. D., 113
Uzoigwe, G. N., 108

Vapñarsky, C. A., 116
Vennetier, P., 110, 111
Victoria, E., 141
Vincent, J., 82
Visaria, P., 71
Von Boventer, E., 177, 188

Webber, M. J., 87
Weeks, J. F., 83
Weiner, M., 76
Whitelaw, W. E., 71
Williamson, J. G., 113–122 *passim*, 135–136, 141, 177
Wilson, G., 143
Wingo, L., 178, 182, 183, 184, 189
Withington, W. A., 70
Wood, G. D., 24, 33, 34
Wood, L. J., 108

Zachariah, K. C., 62

Subject Index

Ābādān, Iran, 160, 164
Abeokuta, Nigeria, 87, 108
Accra, Nigeria, 71, 156
Africanization, 95–96
Agrarian reform, India, 21–45 *passim*
    Peru, 58–59
Agriculture, credit, 21–45 *passim*
    services centres and, 96–97, 157–158
    urban development and, 190
Ahvāz, Iran, 160, 164
Akbar's army, 26
Alagoas State, Brazil, 135
Arāk, Iran, 160, 164
Area Redevelopment Administration, 2
Argentina, 118–119, 122, 174–175, 177, 192
Arusha, Tanzania, 102, 104, 160, 163
Arusha Declaration, 163
Asansol, India, 156
Aurangabad, India, 185
Australia, 115, 174, 175
Austria, 115–119 *passim*

Bahia State, Brazil, 134
Bangkok, Thailand, 60, 61, 65
Barranquilla, Colombia, 186
Basques, 10
Belgium, 174, 175
Bengal State, India, 36, 37
Biafra war, 10, 70
Bihar State, India, 21–45 *passim*
Bogotá, Colombia, 55, 63, 64, 67, 68, 171, 186
Bolivia, 118–119
Boma, 81
Bombay, India, 16, 158, 185

Brahmins, 36
Brasília, 4
Brazil, economic 'miracle', 125–135 *passim*
    health, 134–135
    income disparities, 114–119 *passim*, 124–135
    incomes and city size, 184
    migration, 51, 171
    mortality, 134–135
    primacy and economic development, 174–175, 177, 192
    regional development, 1, 8, 125–135
    urban diseconomies, 191
Bucaramanga, Colombia, 186
Budapest, Hungary, 170
Buenos Aires, Argentina, 16
Buganda, Uganda, 80
Bugondo, Uganda, 82
'Bullet' trains, 9

Calcutta, India, 16, 48
Cali, Colombia, 186
Cameroun, 85
Campina Grande, Brazil, 131
Canada, 115–119 *passim*, 122, 174, 175
Caruaru, Brazil, 134
Cassa per il Mezzogiorno, 2, 9
Caste system, 21–45 *passim*
Ceará State, Brazil, 134, 135
Central Place theory, 85–89, 105–106
Chile, economic development and primacy, 174, 175, 177, 192
    income disparities, 115–119 *passim*, 122
    migration, 52, 58, 61, 63
China, 174, 175

City size and development, 169–197 *passim*
Ciudad Guayana, Venezuela, 156
Colombia, income disparities, 115–119 *passim*, 122
　incomes and city size, 184–186
　migration, 50, 55, 63–64, 67, 68
　regional development, 1
Commerce in rural areas, East Africa, 80–85, 93–96
　India, 33–34
　transfer of wealth, 99–100
Costa Rica, 174–175, 177
Cuba, 7, 58

Dar es Salaam, Tanzania, 102–105 *passim*, 118, 160, 163
Denmark, 9
Dependency theory, 16–17, 99, 121
Development levels, 174, 175, 177
Diffusion and social structure, 32–36
Diffusion in agriculture, 21–46 *passim*, 96–97
Diffusion theory, 21–24
Dodoma, Tanzania, 5, 102, 160, 163
Dominican Republic, 174–175, 177, 192
Dukas, 81–84
Durgapur, India, 23, 156

East Africa, service centres, 77–111
　transport, 97–99
East African economic community, 163
East Anglia, 35
Ecuador, 171, 174, 175, 177, 182
Education, service centres and, 90, 93
　migration and, 58, 68
Egypt, 174, 175
El Salvador, 174, 175, 177
Eldoret, Kenya, 101, 160, 162
Embu, Kenya, 101, 160, 162
Entropy, 173, 191
Esfahan, Iran, 160
Ethnicity, 10, 65, 69–70, 80–85, 95–96

Family planning, 7, 21–45
Fertility decline, 56–57
Finland, 115–119 *passim*, 174, 175

Forbesganj, India, 40
Fortaleza, Brazil, 131–132
France, 2, 115–119 *passim*, 122, 174, 175

Ganges river, 36–37
Geography of Development, 5, 150, 153
Ghana, city system, 86–87
　income disparities, 118–119
　migration, 52, 60, 67, 71
Greece, 115–119 *passim*, 174, 175
Green belts, 2
Green revolution, 7, 37
Growth centres, 5, 100–105, 143–167
Guatemala, 174–177 *passim*
Guayana region, Venezuela, 4, 11, 156
Guayaquil, Ecuador, 171, 192
Gujarat State, India, 40

Haryana, India, 29
Hausaland, Nigeria, 80
Health, 21–45 *passim*, 90–93, 134–135
Honduras, 174, 175, 177
Hungary, 174, 175

Ibos; 80
Ijebu Province, Nigeria, 87
Inca empire, 152
Income disparities, 10, 113–141
　economic development and, 113–122, 135–136
　income convergence, 119–122
　policy objectives and, 122–125
India, agriculture, 7, 15, 21–45
　family planning, 7, 15, 21–45
　income disparities, 115–119 *passim*
　incomes and city size, 122, 184–185
　intermediate urbanization, 34–35
　mutiny, 37
　planning, 1, 5, 7, 21–45 *passim*, 154
Industry, decentralization, 125–126
　development, 103–105, 155–157, 164
　location, 187, 188
Infant mortality, 134
Internal colonialism, 5
International Labour Office (ILO), 25
Iran, 146, 159–160, 163–164

Ireland, 10, 115–119 *passim*
Israel, 52, 174–175
Italy, 2, 8, 9, 115–119 *passim*, 122, 174, 175

Jahuel, Chile, 171
Jalgoan, India, 185
Japan, 9, 115–119 *passim*, 174, 175, 182
Jinja, Uganda, 96, 143, 155
João Pessao, Brazil, 131
Juazeiro do Norte, Brazil, 134

Kakmega, Kenya, 101, 160, 162
Kampala, Uganda, 79, 143
Kelantun, Malaysia, 157
Kenya, central places, 87–89
  growth centres, 5, 100, 146, 159–162
  income disparities, 118–119
  migration, 50, 71
  planning, 77, 86, 88–90, 162–163
Kerala, India, 37, 43
Kilimanjaro Massif, 163
Kinshasa, Zaire, 77
Kisumu, Kenya, 101, 160, 162
Kitgum, Uganda, 85
Kosi irrigation scheme, India, 29–36

Lagos, Nigeria, 77
Lake Victoria, 162, 163
Lal Bahadur Sastri, 40
Lango, Uganda, 80
Libya, 174, 175
Lima, Peru, 65, 192
London, England, 2, 3, 8, 80, 183

Maharastra, India, 29
Malagasy Republic, 93, 96–97, 99
Malawi, 92
Malaysia, 70, 154, 155
Manila, Philippines, 117
Manizales, Colombia, 186
Mashhad, Iran, 160, 164
Mau Mau campaign, 84
Mbeya, Tanzania, 102, 160
Mengo, Uganda, 87, 93, 108
Mexico, income disparities, 118–119
  incomes and city size, 182

migration 53, 63
primacy, 174–177 *passim*, 180, 192
regional development, 1
Mexico City, 16, 63
Migration, age of migrants, 55–59
  impact on departure areas, 54, 70–71
  impact on migrants, 54, 66–68
  impact on urban areas, 48–49, 54, 68–70
  motives of migrants, 50–51, 53, 66–68
  orientation of research, 47–49
  sexual composition, 54–55, 59–61
  skill levels, 61–65
  sources of data, 49–53
  unemployment and, 67–68, 171
Milan, Italy, 9
Mindanao, Philippines, 70
Modernization surfaces, 5
Mombasa, Kenya, 101, 160, 161
Monterrey, Mexico, 53, 63, 64
Morocco, 37, 174, 175
Morogoro, Tanzania, 102–104, 160
Moshi, Tanzania, 102, 104, 160, 163
Mossoro, Brazil, 134
Mtwara, Tanzania, 102, 160
Mukihayaa, 32
Musahars, India, 34
Muzaffarpur, India, 21–45 *passim*
Mysore State, India, 29

Nagpur, India, 185
Nairobi, Kenya, 71, 77, 101, 117
Nakuru, Kenya, 101, 160, 162
Nanded, India, 185
Natal, Brazil, 131, 132
Nationalism, 10
National planning, difficulties facing, 6–8
  evolution, 153–155
  foreign influence, 10–16
  growth centres, 153–158
Netherlands, 115–119 *passim*, 122, 174–175
New Delhi, India, 21, 25, 26, 29
'New towns', 2, 4, 9
New York City, U.S.A., 3, 8
New Zealand, 115–119 *passim*, 174, 175
Nicaragua, 174–175, 177

Nigeria, 5, 80, 87
Norway, 115–119 *passim*, 122, 174, 175
Nyeri, Kenya, 101, 160, 162

Orissa State, India, 37
Overplanning, 7
Owen Falls scheme, Uganda, 143

Pakistan, 154, 157
Panama, 174, 175, 177
Panchayat, 29, 32, 33
Paraguay, 174, 175, 176, 177
Paraíba State, Brazil, 135
Paraná State, Brazil, 135
Paris, France, 3, 80
Pasto, Colombia, 186
Patna, India, 26
Pernambuco State, Brazil, 135
Peru, 60–61, 65, 118–119, 174, 175, 177, 192
Petroleum, 9
Philippines, 118, 174, 175
Piauí State, Brazil, 134
Pinchincha, Ecuador, 171
Poland, 174, 175
Politics, attitudes of migrants, 69
    regional development and, 11
Poona, India, 185
Portugal, 174, 175
Prague, Czechoslavakia, 170
Pseudo-planning, 7
Primacy, 78, 79, 85–89, 173–181
Puerto Rico, 115–119 *passim*
Punjab, India, 29
Puno, Peru, 65
Purnea District, India, 21–45 *passim*

Quebec, Canada, 10
Quito, Ecuador, 171

Racial problems and structure, 10, 65, 69–70, 80–85, 95–96
Railways, 163–164
Rank-size rule, 78–79, 173–192 *passim*
Recife, Brazil, 131, 132, 134, 184
Regional planning, academic interest in, 5–6
    difficulties facing, 6–8

evolution in third world, 1–6, 113–114
experience of developed countries, 8–9
experience of less-developed countries, 10–16, 100–105, 125–135, 159–163
role of growth centres, 146–164
transfer of strategies, 10–16
Rio de Janeiro, Brazil, 16
Rio Grande do Norte State, Brazil, 135
Rio Grande do Sul State, Brazil, 135
Roads, 97–99, 178
Rome, Italy, 9
Rural development, 21–45 *passim*, 70–71, 89–107, 157–158

Saharsa District, India, 29
Salvador, Brazil, 131
Santa Catarina State, Brazil, 135
Santal Parganas, 36
Santhals, 36, 37, 43
Santiago, Chile, 61
São Luis State, Brazil, 131
São Paulo, Brazil, 16, 135, 171, 177, 190
Scotland, 9, 10
Sergipe State, Brazil, 134
Shīrāz, Iran, 160, 164
Sholapur, India, 185
Sierra Leone, 57–58
Sobral, Brazil, 134
Social Structure, effect on diffusion, 21–45 *passim*
    migration and, 56–58
South Korea, 174, 175
South Wales, 9
Spain, 115–119 *passim*, 174, 175
Sri Lanka, 154
Sudene, 4, 16, 125, 126, 128
Sweden, 9, 115–119 *passim*, 122, 174, 175
Switzerland, 174, 175

Tabora, Tanzania, 102, 160
Tabriz, Iran, 160, 164
Tanga, Tanzania, 102, 104, 160, 163
Tanzania, growth centres, 5, 100–105, 146, 159–160, 163–164

income disparities, 118–119
national plans, 78, 90, 155, 163–164
rural transport, 98
Tan-Zam railway, 163–164
Tehran, Iran, 160, 164
Tema, Ghana, 156
Tennessee Valley Authority (TVA), 1, 2
Teso District, Uganda, 78, 87, 91–94, 97
Thailand, income disparities, 118–119
migration, 52, 60, 61
primacy, 174, 175
Thika, Kenya, 101, 160, 162
Tokyo, Japan, 9, 182
Tolas, 38
Transport, 97–99, 150–151, 163–164, 178
Trengganu, Malaysia, 157
Trinidad, 154, 155
Turin, Italy, 9
Turkey, 174, 175

Uganda, education, 93
health, 90–93
income disparities, 118–119
primacy, 78–79
rural commerce, 93–95
rural transport, 97–99
service centres, 77–112 *passim*
Unemployment, 67–68, 83, 171
United Kingdom, income disparities, 115–119 *passim*
primacy and development, 174, 175
regional development, 2, 3, 9
United States of America, income disparities, 115–120 *passim*, 122
income levels and city size, 184–185
primacy, 174, 175, 192
regional development, 1–2

Urban development, agglomeration economies, 177–179, 182–183
city-size distribution, 78–79, 85–89, 172–181
income and city size, 182–187
infrastructure costs, 69, 85, 182–183, 187–188
intermediate urbanization, 34–35, 81, 96–97, 157–158
internal city structure, 81–85, 171
migrants in urban areas, 48–49, 54, 68–70
pre-industrial cities, 169
relations with rural areas, 89–105, 157–158
relationship with economic development, 89–105, 169–192 *passim*
urban diseconomies, 3, 8–9, 182–183, 187–188, 190–191
Urban planning, 8–9, 83–84, 88, 169–196 *passim*
Uruguay, 174–175, 177, 192

Venezuela, 4, 11, 56, 118–119, 174–175, 177, 192
Visayan, Philippines, 70

Wales, 9, 10
Warsaw, Poland, 170
West Germany, 115–119 *passim*, 122, 174, 175
World health organization, 25

Yugoslavia, 174, 175

Zambia, 94
Zamindar, 34–43